Current Issues in General, Organic, and Biochemistry

Current Issues in General, Organic, and Biochemistry

Selected Readings

Edited by
Michael R. Slabaugh
Weber State University

West Publishing Company
Minneapolis/St. Paul ◆ New York ◆ Los Angeles ◆ San Francisco

Design and production management by Judith Mara Riotto
Cover image credited to Robert Daemmrich, Tony Stone Images

WEST'S COMMITMENT TO THE ENVIRONMENT

In 1906, West Publishing Company began recycling materials left over from the production of books. This began a tradition of efficient and responsible use of resources. Today, 100% of our legal bound volumes are printed on acid-free, recycled paper consisting of 50% new paper pulp and 50% paper that has undergone a de-inking process. We also use vegetable-based inks to print all of our books. West recycles nearly 27,700,000 pounds of scrap paper annually—the equivalent of 229,300 trees. Since the 1960s, West has devised ways to capture and recycle waste inks, solvents, oils, and vapors created in the printing process. We also recycle plastics of all kinds, wood, glass, corrugated cardboard, and batteries, and have eliminated the use of polystyrene book packaging. We at West are proud of the longevity and the scope of our commitment to the environment.

West pocket parts and advance sheets are printed on recyclable paper and can be collected and recycled with newspapers. Staples do not have to be removed. Bound volumes can be recycled after removing the cover.

Production, Prepress, Printing and Binding by West Publishing Company.

COPYRIGHT © 1996 by   WEST PUBLISHING CO.
610 Opperman Drive
P.O. Box 64526
St. Paul, MN 55164–0526

All rights reserved
Printed in the United States of America
03 02 01 00 99 98 97 96   8 7 6 5 4 3 2 1 0

ISBN 0–314–07740–5

Contents

Preface / ix

1. **Metrics: Mismeasuring Consumer Demand**
   by Michael Chapman
   (*Consumers' Research*, February 1994) / **1**

2. **Going Metric: American Foods and Drugs Measure Up**
   by Judith Randal
   (*FDA Consumer*, September 1994) / **5**

3. **Serendipity**
   by J.E. Oldfield
   The development of uses for selenium has been marked by innovative conceptions, misinterpretations, and just plain accidents.
   (*CHEMTECH*, March 1995) / **8**

4. **Calcium: Here's How to Bone Up on This Essential Mineral**
   (*Mayo Clinic Health Letter*, May 1995) / **13**

5. **Gold: Noblest of the Nobles**
   by Ken Reese
   Imagine, if you will, a prospective investor being told that "Gold, atomic number 79, is a third-row transition metal in Group IB of the periodic table . . ."
   (*Today's Chemist at Work*, October 1994) / **16**

6. **Heavy-Ion Research Institute Explores Limits of Periodic Table**
   by Michael Freemantle
   German scientists are confident that element 112 will soon be created, followed eventually by elements 113 and 114.
   (*Chemical & Engineering News*, March 13, 1995) / **19**

7. **The New Miracle Drug May Be—Smog?**
   by Randi Hutter Epstein
   Nitric oxide figures in many diseases, and possible cures.
   (*Business Week*, December 5, 1994) / **23**

8. **The Big Squeeze in the Lab: How Extreme Pressure Is Creating Exotic Materials**
   by Ruth Coxeter, with Peter Coy
   (*Business Week*, February 13, 1995) / **25**

9. **Chemical Techniques Help Conserve Artifacts Raised from *Titanic* Wreck**
   by Michael Freemantle
   Compexing agents and electrolysis restore some artifacts from the Titanic, *providing insights for research in deep-sea areas.*
   (*Chemical & Engineering News*, October 17, 1994) / **27**

10. **Carbon Dioxide: Global Problem and Global Resource**
    by Kimberly A. Magrini and David Boron
    Global emissions of carbon dioxide are of increasing environmental concern. The problem could be eased by using the gas as a resource in chemical processes.
    (*Chemistry & Industry*, December 19, 1994) / **31**

11. **Physicists Create New State of Matter**
    by Gary Taubes
    By cooling a crowd of atoms to within a hair of absolute zero, researchers have made the crowd behave as one, opening a new arena for physics.
    (*Science*, July 14, 1995) / **36**

12. **Tapping the Market for Bottled Baby Water**
    (*Tufts University Diet & Nutrition Letter*, March 1995) / **39**

13. **Gas Hydrates Eyed as Future Energy Source**
    by Ron Dagani
    (*Chemical & Engineering News*, March 6, 1995) / **41**

14. **Dances with Molecules: Controlling Chemical Reactions with Laser Light**
    by Richard Lipkin
    (*Science News,* May 27, 1995) / **43**

15. **Complexities of Ozone Loss Continue to Challenge Scientists**
    by Pamela S. Zurer
    Severe Arctic depletion verified, but intricacies of polar stratospheric clouds, midlatitude loss still puzzle researchers.
    (*Chemical & Engineering News,* June 12, 1995) / **46**

16. **Drug Giants Ready to Enter Stomach-Medicine Battle**
    by Steve Sakson
    (Associated Press, July 16, 1995) / **51**

17. **Having the Last Gas**
    by Nigel P. Freestone, Paul S. Phillips, and Ray Hall
    Alternative energy sources are frequently the butt of jokes, but with vast areas of land being claimed for waste burial, landfill gas is making a serious challenge as a potentially important commodity.
    (*Chemistry in Britain,* January 1994) / **53**

18. **Making Molecular Filters More Reactive**
    by Robert F. Service
    (*Science,* September 2, 1994) / **57**

19. **The Chlorine Controversy**
    by Gordon Graff
    Some environmentalists are calling for the total elimination of the many chemicals containing chlorine. While that will not likely happen, government and industry could take measures to reduce the use of unsafe compounds.
    (*Technology Review,* January 1995) / **59**

20. **Wine and Heart Disease**
    by Andrew L. Waterhouse
    Phenolic substances may be the compounds responsible for the reduced incidence of coronary heart disease seen in people who regularly consume wine.
    (*Chemistry & Industry,* May 1, 1995) / **64**

21. **Amoco, Haldor Topsoe Develop Dimethyl Ether as Alternative Diesel Fuel**
    by A. Maureen Rouhi
    Dimethyl ether use in diesel engines cuts soot and nitrogen oxides emissions, reduces engine noise.
    (*Chemical & Engineering News,* May 29, 1995) / **69**

22. **Clearing the Air**
    (*University of California at Berkeley Wellness Letter,* October 1994) / **72**

23. **Pain, Pain Go Away**
    by Ruth Papazian
    (*FDA Consumer,* January–February 1995) / **74**

24. **Drugstore Deceptions: Separating Hype from Hope**
    Many self-care remedies offer only empty promises. They're a waste of money—and some are risky as well.
    (*Consumer Reports on Health,* March 1995) / **77**

25. **Caffeine: Grounds for Concern?**
    (*University of California at Berkeley Wellness Letter,* March 1994) / **80**

26. **Green Chemistry at Work**
    by John Frost
    Products can be made from glucose instead of benzene.
    (*EPA Journal,* Fall 1994) / **83**

27. **What's Wrong with Sugar?**
    Sugar's reputation as a health heavy is much overblown.
    (*Consumer Reports on Health,* October 1994) / **85**

28. **Fiber Bounces Back**
    After all the jokes about oat bran, fiber is getting the last laugh: It does help fight deadly disease.
    (*Consumer Reports on Health,* March 1995) / **88**

29. **New Oils for Old**
    by Denis J. Murphy
    Plant oils are a versatile source of renewable feedstocks for manufacturing a wide variety of industrial products, ranging from pharmaceuticals and cosmetics to lubricants and fine chemicals.
    (*Chemistry in Britain,* April 1995) / **91**

30. **Estrogen: Friend or Foe?**
    by Marilynn Larkin
    (*FDA Consumer,* April 1995) / **95**

31. **Protein Devices May Increase Computer Speed and Memory**
    by Michael Freemantle
    Prototype devices use bacterial protein memory cubes to store up to 300 times more information than current devices.
    (*Chemical & Engineering News,* May 22, 1995) / **99**

32. **Designer Proteins: Building Machines of Life from Scratch**
by Richard Lipkin
(*Science News*, December 10, 1994) / **102**

33. **Novo Nordisk's Mean Green Machine**
by Julia Flynn, with Zachary Schiller, John Carey, and Ruth Coxeter
It's at the top of a growing market: Finding natural substances to replace chemicals.
(*Business Week*, November 14, 1994) / **106**

34. **Better by Design: Biocatalysts for the Future**
by Alan Wiseman
Useful industrial biocatalysts are currently being redesigned by using biological and chemical techniques, to provide novel enzymes and mimics for use in food processing and pharmaceutical manufacture.
(*Chemistry in Britain*, July 1994) / **109**

35. **Life in Boiling Water**
by Robert M. Kelly, John A. Baross, and Michael W.W. Adams
Apart from pandering to our scientific curiosity, studies of microbial lifeforms that can exist at temperatures above 100 °C are revealing a host of commercially important enzymes.
(*Chemistry in Britain*, July 1994) / **113**

36. **The Gene Kings**
by John Carey, with Joan O'C. Hamilton, Julia Flynn, and Geoffrey Smith
Two scientists are changing how DNA is mined—and drugs are discovered.
(*Business Week*, May 8, 1995) / **117**

37. **Environmental Risks in Agricultural Biotechnology**
by Peter Kareiva and John Stark
Biotechnology has given rise to many fears within the public domain. Basic ecological experiments should distinguish the real dangers from the imaginary.
(*Chemistry & Industry*, January 17, 1994) / **122**

38. **Transgenic Livestock in Agriculture and Medicine**
by Caird E. Rexroad, Jr.
Although still in its relative infancy, transgenics is already offering a chance to aid organ transplants and produce drugs, as well as enhancing foodstuffs.
(*Chemistry & Industry*, May 15, 1995) / **127**

39. **Cancer-fighting Foods: Green Revolution**
by Kristine Napier
(*Harvard Health Letter*, April 1995 supplement) / **131**

40. **B Makes the Grade**
A flurry of recent studies is propelling the once-lowly B vitamins toward the head of the class.
(*Consumer Reports on Health*, June 1995) / **135**

41. **Hypoglycemia: Fact and Fiction about Blood Sugar**
by Jamie Spencer
(*Harvard Health Letter*, April 1995 supplement) / **138**

42. **Marketers Milk Misconceptions on Lactose Intolerance**
(*Tufts University Diet & Nutrition Letter*, December 1994) / **141**

43. **Nutrition: Pasta Is Not Poison**
by Kristine Napier
(*Harvard Health Letter*, July 1995) / **145**

44. **The New Skinny on Fat**
by Traci Watson
A protein that makes mice shed pounds might ultimately help people, too.
(*U.S. News & World Report*, August 7, 1995) / **148**

45. **Cutting Cholesterol: More Vital Than Ever**
(*Consumer Reports on Health*, February 1995) / **151**

46. **Water: The Ultimate Nutrient**
by Nancy Clark
(*The Physician and Sportsmedicine*, May 1995) / **154**

47. **Iron: Too Much of a Good Thing?**
Once touted as a pep tonic, iron may actually increase the risk of deadly disease.
(*Consumer Reports on Health*, July 1994) / **156**

48. **Artificial Blood May Be a Heartbeat Away**
by Ron Stodghill II
Northfield Labs is close to perfecting recycled hemoglobin.
(*Business Week*, June 5, 1995) / **159**

Answers / 161

Preface

Chemistry is an exciting and dynamic field. Researchers in chemistry are making discoveries almost every day that meet our society's needs and improve our quality of life. Chemistry plays a crucial role in attempts to discover new processes, develop new sources of energy, produce new products and materials, provide more food, and ensure better health. Unfortunately, students rarely see this view of chemistry. It is far more common for students to describe chemistry as difficult, dangerous, boring, required, smelly, hazardous, toxic, or dull. This negative image is perpetuated by the mass media as they describe such newsworthy topics as toxic chemical spills, hazardous waste disposal problems, and nerve gas attacks.

I hope this readings supplement will help students gain a more balanced understanding of the true role of chemistry in our society. This collection of forty-eight articles was gathered from general-interest and science magazines. Each article was carefully chosen to provide an up-to-date supplement to the material students might encounter in a first-year chemistry class. Each article begins with a brief overview and introduction of the topics covered in the selection. Following each article are a few questions to help students focus on the main points of the discussion.

Acknowledgments

The editor and West Publishing wish to express their sincere thanks to the many magazines and journals that allowed us to reprint their articles. I am also indebted to three university science majors: Michael C. Slabaugh, who assisted in the library research and in typing portions of the manuscript, and Mark and Christina Slabaugh, for an excellent job writing many of the end-of-article questions.

Very few of the scientific and technological achievements taken for granted today could have been accomplished without careful measurements. For more than a century, the nations of the world have recognized the need to use a standard system of measurement. The metric system has a number of advantages compared to other measurement systems and is used by almost every country. However, widespread adoption of the metric system has not yet occurred in the United States. In fact, a recent Gallup Poll showed that 64 percent of the U.S. population opposed metric conversion. Even though Congress passed the Metric Conversion Act and an amendment in 1988 that forces federal agencies to require metric units, it is still uncertain whether the public will embrace the metric system.

Metrics: Mismeasuring Consumer Demand

Michael Chapman

Is the Department of Justice, the top governmental agency responsible for law enforcement, violating the law? Apparently, yes. The law in question concerns the use of the metric system of weight and measure as enacted by the Metric Conversion Act of 1975 and amended in the Omnibus Trade and Competitiveness Act of 1988. This amendment on metric usage (Public Law 100-418, Section 5164) "declares that the policy of the nation is to designate the metric system as the preferred system of measurement for trade and commerce, and requires each federal agency to use metric units in all or as many of its procurements, grants, and other business-related transactions as is economically feasible by the end of fiscal year 1992." According to a report by the Congressional Research Service at the end of fiscal year 1992, the Justice Department "does not appear to be complying with the [metric usage] law."

The Justice Department is not alone.

The latest information available indicates that no less than 22 of 37 federal agencies have either completely or partially failed to comply with the metric conversion law. These "outlaw" agencies include: Department of Education, Department of Transportation, Federal Trade Commission, U.S. Postal Service, General Services Administration, and the Government Printing Office. Despite these apparent violators of federal law, don't expect the U.S. government to indict itself. Attempts at metric conversion in the private sector never really got off the ground. It is still uncertain whether this attempt at conversion in the public sector will survive, let alone succeed.

Metric History Lesson
The attempt to replace the English (or customary) system of weight and measure, which is based on inch/pound/quart measurements, with the metric system, which is decimal-based and uses meters, grams, and liters for measurement, has a long history.

The metric system was born during the French Revolution. In the United States, both Thomas Jefferson and John Quincy Adams advocated, unsuccessfully, metric conversion. By an Act of Congress in 1866, metric usage was legalized in the United States on a voluntary basis. In 1875, along with 17 other nations, the United States signed the Treaty of the Meter. This agreement established the International Bureau of Weights and Measures in Sèvres, France, to provide metric standards of measurement for worldwide use. These standards for length and mass were adopted in the United States in 1893. In 1960 the metric standards were revised. This modernized version of the metric system is known as *Le Systeme International d'Unites* (International System of Units) or SI. Metrics have been legal on a voluntary basis for more than 100 years; but except in those fields that are metric-dependent—science and trade—widespread metric conversion in the United States has not occurred.

Reprinted with permission from *Consumers' Research*, 800 Maryland Ave., N.E., Washington, DC 20002. Copyright 1994, Consumers' Research, Inc.

To promote metric conversion in the United States, Congress passed the Metric Conversion Act of 1975. This Act called for a voluntary conversion by individual groups and industries. However, this attempt failed. Americans, by and large, rejected the system. "The switch to metric was perceived as hostile to consumers," said *Government Executive* in 1990. "The public objected loudly to road signs showing distances in kilometers, to temperatures in Celsius, and to gasoline sold in liters."

In assessing this unsuccessful attempt at metric conversion, G.T. Underwood, former director of the Office of Metric Programs at the Department of Commerce, says: "Arguments about lost export markets got mixed up with the need for metric road signs. The general public resented what seemed an unnecessary social nuisance. Most U.S. firms, seeking not to aggravate U.S. customers, didn't change their products, the ostriches prevailed, and the movement essentially stalled." On a related note, a General Accounting Office (GAO) report in 1978 found: the total cost of metric conversion was indeterminable but substantial, somewhere in the billions-of-dollars ranges; conversion would result in higher consumer prices and reduced U.S. productivity; U.S. and world trade would not be hampered by a dual system of English and metric measurement; and, there was no evidence that a solely metric system would benefit the U.S. economy.

As a result of the reaction in the 1970s, overt enthusiasm for metric conversion in the public and private sector waned—until 1988.

As mentioned, in the Omnibus Trade and Competitiveness Act of 1988, the metric system is designated the "preferred system of measurement" in the United States and requires all federal agencies to "go" metric. In 1991, President George Bush issued Executive Order 12770, which clarified the role of the Commerce Department to direct and coordinate all federal agencies in converting to the metric system. With these directives, proponents of metrication plan to stimulate conversion in the United States from the top down—from government, to industries, to small businesses, and eventually to consumers.

"The amended Metric Conversion Act of 1975 and the 1991 Executive Order provide both the rationale and the mandate for a transition to the use of metric units," says Dr. Gary P. Carver, chief of the Metric Program at the National Institute of Standards and Technology (NIST). Metric conversion in the federal government could finally tip "the scale toward general acceptance of the metric system," notes *Governmental Executive*.

But how exactly will this latest attempt at metric conversion work? And how will it affect consumers?

The Case for Metrication

Stimulate conversion. The new attempt at metric conversion strongly encourages American industries that sell products to the government to produce these products in metric units. Conversion is voluntary for private industry. But if an industry sells to a government that is required by law to purchase metric-sized products, then what will that industry do? Answer: Either stop selling to that government or convert its products to metric measurement. Carver says the main objective of metric conversion in the federal agencies is to "stimulate" people, not force them to convert to metrics. The budgetary power of the 37 federal agencies involved is a powerful stimulus.

The General Services Administration (GSA), for instance, spends more than $2 billion a year on procurement. The entire federal bureaucracy spends more that $300 billion a year on goods and supplies. Under the law, all federal agencies must use the metric system in their procurements, grants, and other business-related activities. More that $300 billion in procurements will unquestionably affect the nation. Hence, industries have converted or are in the process of converting to the metric system.

Automobiles built by General Motors, Ford, and Chrysler are constructed using metric measurements, as are computer designs made by Xerox and IBM. Lockheed and Boeing aircraft have converted to remain eligible for Pentagon contracts. Soft drinks, liquor, tires, film, cameras, skis, and many weapons systems are either produced, sold, or labeled in metric measurement. The now-stalled Strategic Defense Initiative was built according to metric standards. And, by February 14, 1994, all consumer product labeling and packaging must be in both English and metric measurements. By "stimulating" businesses to convert either fully or partially to metric usage, metric proponents hope that Americans will eventually accept metrication as more and more consumer products and services are "metricised."

"We made a mistake after 1975 by trying to force metrics down peoples throats," says Underwood. "This time, business is leading the way, and social and cultural change will follow." The metric system is apparently making its long march through the governmental institutions.

Trade and jobs. Metric proponents say that conversion to the metric system is necessary and inevitable. Most of the world uses metric measurement and international trade involves metric-sized products. If American industry wants to stay competitive in the global marketplace, the reasoning goes, then U.S. industry better get on the metric bandwagon. By going metric the U.S. government "would open the door for new markets and thereby help to create the new jobs this nation so drastically needs," says Senator Claiborne Pell (D-R.I.). "[I]t is time for our government to assume a leadership position on the metric issue, instead of passively waiting for market forces to reverse our archaic system of measurement." The Commerce Department estimates that U.S. exports could be increased by up to 20% by offering metric-sized goods to international markets.

In speech before the National Metric Conference in 1992, President Bush endorsed metrication of U.S. products. President Clinton also supports metrication: "All developed nations except the United States use the metric system, and it is clear that our country will benefit by encouraging voluntary metric use by industry. These efforts can enhance America's competitive edge and help create new jobs and opportunities for our people." The European Community, which has a buying public of 320 million people, threatened to bar the importation of non-metric products after December 1992, but this deadline has been extended to December 1999.

Metrication of U.S. industry, say its proponents, will lead to better trade with Canada, Mexico, Europe, and the nations of the Pacific Rim. "Adopting metric is only one key to seizing these opportunities, but an important one that, when combined with other 'attitude adjustments,' will greatly affect the economic health of this country and our future standard of living," said Underwood.

Other benefits. In addition to its effect on U.S. exports, metric conversion will benefit the average consumer, says NIST. Metrication should promote standardized and simpler product packaging, which will reduce the number of package sizes, simplify price comparisons, and lower packaging and shipping costs. These savings will reach the consumer, says NIST. In switching to metric, the U.S. liquor industry reduced the number of its container sizes from 53 to seven, which resulted in a substantial savings in production costs. In its metric conversion, IBM reduced 38,000 part numbers in fasteners to 4,000.

The Case against Metrication

Regardless of the benefits of metric conversion to U.S. trade, opponents of metrication say there is no need for the United States to switch systems to accommodate the rest of the world.

"The people of this country should not be coerced to convert to the 200-year-old, artificially contrived metric system. Metrics are a language of technocracy and multinational trade. Let science and industry use the metric system as they need it," says Seaver Leslie, head of the Americans for Customary Weight and Measure, a not-for-profit group dedicated to retaining the English system of weight and measure.

Costs. A survey by the National Federation of Independent Business (NFIB) in 1979 found that 69% of 55,401 of its members surveyed opposed metric conversion because of costs. "Metric conversion benefits large, manufacturing industries and most of these are already undergoing conversion, but the metric system should not be forced down the throats of all businesses in America. The cost to small firms, in time lost and wasted materials, could never be recouped," the NFIB said at the time, Fourteen years later, the NFIB, with 610,000 members, has not changed its position. "Most small businesses are opposed to metric conversion because of costs," says Terry Hill, a spokesman for the group. On a related note, the Nuclear Regulatory Commission estimates that it will cost the agency $2 to $3 million to convert to metrics. As mentioned, the GAO reported that the total cost of metric conversion of the United States was indeterminable—but would be in the many billions of dollars.

Confusion. There is a single, world standard for the inch. But unknown to most people there are various metric systems in use today. The SI system proposed for the United States "is materially different from the metric system of other nations, [and] there is much evidence that these nations intend to protect their interests and thus are reluctant to adopt SI in its entirety. Even if the United States converts to SI . . . still no single worldwide system of measurement would exist," according to the GAO.

Metric conversion from the top down, if successful, would eventually affect nearly all aspects of daily life. Workers would have to be retrained, tools replaced, machinery modified, map distances changed, etc. Food and clothing sizes would change. Everything. Out centuries old way of doing things (and thinking about them) would change.

Highway signs. The 1978 Federal Aid Highway Act prohibited the use of federal funds for metric-only signs. This part of the law was overturned when the Intermodal Surface Transportation Efficiency Act passed in 1991. As a result of the 1988 metric usage law, all highway and highway-related construction funded by the federal government will be done in metric measurement. The deadline for this conversion is September 30, 1996. Although Americans objected to metric road signs when they were proposed in the 1970s, the Department of Transportation (DOT) is currently reviewing comments about metric conversion of highway signs. (As of this printing, it had not made a decision about sign conversion). In previous responses to this issue, 47% of states told the Federal Highway Administration they opposed metric conversion and only 18% supported it. Recent reports on the response to metric conversion of highway signs suggest that only a few states oppose the conversion. However, Amy Steiner of the American Association of State

Highway and Transportation Officials says that states don't want to spend money on highway-sign conversion. "They [states] fear citizen backlash. Citizens don't want metric thrown up at them," she says.

At least one congressman isn't prepared to wait for the Transportation Department to make a decision. After introducing a bill (H.R. 3343) to prohibit the expenditure of federal funds on metric system highway signing, Representative Pat Williams (D-Mont.) said: "Changing over some areas in our daily lives to metric may make sense in some areas. However, modifying our highway signs does nothing to promote international trade. It does nothing to keep businesses in America. It will cause confusion. Some estimates peg the national cost to converting the nation's highway signs at more than $200 million." To date, no action on this bill has occurred.

Consumer fraud. Metrication would confuse consumers and probably encourage consumer fraud. "Consumers would not know whether they are getting their money's worth for things sold by length, volume, or weight. They may not be able to recognize price increases," said the GAO. For instance, a gallon of gasoline that costs $1.21 would cost 32 cents per liter (one gallon equals 3.8 liters). "Gas guzzler taxes and registration fees based on vehicle weight are other areas for abuse," says the National Motorists Association of Dane, Wisconsin. The tables based on the metric system are different and costlier than existing tables, which are based on the English system. "When the wine and liquor industry changed the half-gallon to a 1.75 liter bottle, a 7½% decrease in volume occurred with no proportionate decrease in price," says Leslie.

Who Really Wants Metrics?
As mentioned, voluntary usage of the metric system was legalized in 1866. But Americans don't seem to want the system. A 1991 Gallup Poll showed that 64% of the U.S. population opposed metric conversion. Apparently some government suppliers are having trouble converting to metric precisely because Americans still don't want metric products. "Companies tell us that they're not going to change until their customers demand it," says Carver. "It's tough to get the Department of Commerce to switch to A4 [metric-sized] paper," he says. Other agencies cite similar problems.

To "metrify" to a large extent, the Postal Service, according to the GAO, said that "it would have to convince its vendors and customers to do so." However, many of these clients do not conduct business on an international scale. As a consequence, "when the Postal Service buys equipment that was designed in metric dimensions, it still has to convert some parts back to inches to ensure a ready and economical parts supply," said the GAO. The GSA said "[I]t can encourage its suppliers to convert to the metric system but cannot dictate to them."

If the Commerce Department, the agency responsible for directing metric conversion among the 37 federal agencies, is finding it difficult to switch to metric-sized paper, then the future of total metric conversion in the government seems dubious. (Nonetheless, Carver remains optimistic and cites the success of the highway transition plan for 1996.)

As we go to press, the GAO had not released its update on metric conversion. However, indications suggest that metrication of federal agencies has not proceeded at the pace and to the extent its planners had envisioned back in 1988. As William Freeborne, the metric coordinator for the Department of Housing and Urban Development, says: "We're not in great shape." Even Carver says he is not comfortable with the latest report.

An interesting point is that taxpayers, who have repeatedly expressed their rejection of metrics, have been paying for forced government conversion, even though use of the metric system on a voluntary basis was legalized in 1866—128 years ago. ∎

Questions

1. Why was the International Bureau of Weights and Measures established?

2. What action did President Bush take in 1991 to encourage conversion to the metric system?

3. List two benefits for the United States of converting to the metric system.

Answers begin on page 161.

2 *Ever since 1866 when President Andrew Johnson signed an act making metric measurements legal, the United States has been on a long and slow road to converting from a pound-inch (English) system of measurement to the metric system. Many people feel that our reluctance to incorporate the metric system puts us at a great disadvantage in international trade. Today, the global marketplace with its instantaneous communication requires a common, standard set of measurements worldwide. Virtually all of the countries of the world, including England and all other English-speaking nations, have adopted the metric system. Two fields in which rapid progress is being made in metrication in the United States are foods and drugs. The U.S. Food and Drug Administration seems committed to this effort.*

Going Metric: American Foods and Drugs Measure Up

Judith Randal

Years ago, nearly every principality, dukedom and kingdom had its own set of weights and measures that often differed wildly from standards in effect a few miles away. That worked well enough for small, self-contained societies. But as the horizons of commerce broadened, local measurement units no longer sufficed.

Today, in a time of rapid transport and instantaneous communication, there is only one marketplace and it is global. Trading in it requires a common, standard set of measurements worldwide. We have such a set of standards. It is popularly known as the metric system and more formally as the International System of Units (SI).

Until now, the United States has been very much like the Boy Scout in the Labor Day parade whose proud mother exclaimed, "Look, they are all out of step but Johnny." Except for two minuscule players in the global marketplace—Liberia and Myanmar (Burma)—ours has been the only trading nation not routinely conducting its commercial affairs in metric units.

This has been a source of problems: If we sell rice or flour by the pound, how does this sit with a purchaser in some foreign country who is used to buying rice or flour by the kilogram? And what does it do to our competitive position if U.S. products are the only ones on a store shelf in a distant land labeled in weight units that make no sense to would-be purchasers?

Through a 1992 amendment to the 1967 Fair Packaging and Labeling Act, Congress has ordered that the packaging of many consumer products display both the customary inch-pound and metric designations. Final regulations for products in the Food and Drug Administration's domain—most processed foods, virtually all nonprescription drugs, and personal care items ranging from liquid makeup to sunscreens—are expected soon.

Once the regulations are final, manufacturers of the products will have a grace period to use up any labels already on hand that disclose only inch-pound system measurements. And some of the hundreds and thousands of products in FDA's jurisdiction—adhesive bandage strips are an example—will not be subject to the new requirement, at least not immediately. Meanwhile, because some U.S. companies have realized the benefits of "going metric," their products are already labeled in both metric and traditional units. Typical is this notation on a can of condensed tomato soup: "10 3/4 oz.— 305 grams." And the labels on prescription drugs have had metric designations for many years.

Switch Is On
Since mid-century, all other English-speaking nations have made the switch to metric. For example, both Canada and Australia, longtime adherents to the English system of measurement, are now comfortably in the metric camp—as indeed is England itself. Conversion to metric in other English-speaking countries was not done overnight or

Reprinted from *FDA Consumer*, September 1994, pp. 23–26.

without public information and education campaigns. Starting in 1970, Australia phased in new measurements one at a time with a series of "M-days" ("M" for metric), each preceded by a barrage of publicity through the news media. As each deadline passed, an old standard of weight or volume or length or area disappeared and a metric one took its place. Canadians followed a similar course of gradualism over a 10-year period starting in 1973. Today, according to an official at the Canadian embassy in Washington, "older people remember the old measures, but younger people don't know what a gallon is."

If gradualism is the key to success in metric conversion, the American experience should be a cakewalk. We have been on the metric track for almost 120 years; indeed, it is an irony of history that the world's last major holdout was one of the original signatories to the first international metric agreement. This was the Treaty of the Meter, signed in Paris in 1875. The United States was the only English-speaking country among the treaty's 17 original signers.

Even earlier—in 1866—President Andrew Johnson signed an act making metric measurements legal anywhere in the United States. In 1893, metric units were declared to be the country's fundamental standards of measurement. This means, for instance, that the yard is, in effect, a derivative measurement unit equal to 0.9144 meter, and the pound a unit equal to 0.454 kilogram.

In the late 1970s, the American government started a push to metric, but its momentum slowed in the following decade. Even so, some changes were made—for example, the beverage industry switched from "English" measures to their metric counterparts. Most big bottles of soft drinks are now labeled "2 liters." Spirits, once packaged in pints, fifths, quarts, and gallons are now sold in roughly equivalent metric units: 375 milliliters (mL), 750 mL, and the like.

"Nobody's stopped drinking because of the changeover," says Gary Carver, Ph.D., of the Department of Commerce's National Institute of Standards and Technology, who plays a major role in coordinating U.S. government activities in the metric field.

Dual Labeling
Much of what has been done to date has been at the instigation of business, spurred by America's need to better its bargaining position in a growing and increasingly competitive global marketplace. The steps taken by Congress under the Fair Packaging and Labeling Act to ensure that metric gets at least equal billing with the old inch-pound system on a wide variety of foods and other commodities and products will help still more to bring that about.

"Dual labeling" (as the practice of displaying both old and new measurements is known) will probably eventually disappear, leaving metric units the only ones on product labels. For now, however, the use of both systems will give consumers unfamiliar with metric a chance to learn about it. Meanwhile, children in school from the very earliest years—even kindergartners in some school systems—are learning about metric and will be fully up to speed on this system when they, in turn, become consumers in their own right. Indeed, there are many middle-aged Americans today who learned metric when they were in school decades ago, but who may have forgotten the system because they never put it to practical use.

Fortunately for them and for today's shoppers who did not learn metric in school, there are handy approximations between the old, familiar units and metric ones. Although virtually all prescription drugs have long been dispensed and labeled in metric terms, consumers may need some extra help for a few OTC conversions. For example, a 5-grain aspirin tablet is 325 milligrams in metric terms.

Some vitamins are often rated in international units, but IU is a measure of biological activity, not weight or volume, so the congressional mandate for metric labeling doesn't apply. However, vitamin manufacturers that use IU designations are being pushed toward metric labeling by the U.S. Pharmacopeia (U.S.P), a private agency that works closely with FDA. Starting in July, any vitamin product claiming to meet U.S.P. standards and saying so on its label also has to disclose its content in metric terms.

Easy Does It
As anyone who has traveled abroad understands, metric is quite easy to get used to. It's based entirely on the decimal system—everything is divisible or multipliable by factors of 10. For example, a kilogram is 1,000 grams, a liter is 1,000 milliliters, a meter is 1,000 millimeters (or 100 centimeters) and a kilometer is 1,000 meters. Products sold by volume are in liters, those sold by weight are in grams, and those sold by length are in meters (or decimal multiples or fractions of them). No confusion there. Contrast these units with pounds for dry weight (16 ounces), quarts for liquid measure (32 fluid ounces—and fluid ounces aren't the same as dry ounces), yards at 3 feet or 36 inches each, and miles consisting of 1,760 yards or 5,280 feet. You don't have to be a rocket scientist, as the saying goes, to see that metric is actually more consistent than traditional U.S. measures.

And you don't have to be a master chemist to convert cookbook recipes to metric either. For "1 teaspoon" in an old-style recipe, read "5 milliliters (mL)," for "1 tablespoon" read "15 mL," and for "1 cup" of liquid read "250 mL." More and more cookbooks nowadays contain conversion tables—including the difficult and crucial one from Fahrenheit measurement of temperature to Celsius—and measuring cups are available marked in metric on one side and ounces-pint on the other.

Commenting that the metric system is easier and more logical than inch-pound, Commerce's Carver says, "but even if it wasn't we'd have to convert simply to be in step commercially with the rest of the world. Not to do so would be to continue to impose trade barriers on ourselves." With 25 million Canadians to the north and 85 million Mexicans to the south, all allied with us in the North American Free Trade Alliance (NAFTA) and with billions more customers worldwide subject to the General Agreement on Tariffs and Trade (GATT), says Carver, "we *have* to go metric."

Survivors

Old ways die hard, to be sure, and there won't be a "metric police" patrolling the shopping malls, so it is reasonable to expect that some inch-pound units will survive in folk culture even after they disappear from formal commerce. The Australians are simon-pure metrists—except for their dogged adherence to "shouting a schooner" when they visit the local pub, a schooner being a large mug that Australian beer drinkers simply refused to give up when metric came in. Similarly, some older people in Germany refer to a *Pfund* when buying half a kilogram of some bulk product, a *Pfund* being an old measure equivalent to the English pound. Britain, now otherwise solidly metric, still sticks to miles on the highway and, in popular parlance, you still hear some Britons talk of their weight as being so many *stone*. (A stone is the equivalent of 14 pounds.) However, petrol (gasoline) is sold by the liter, spelled litre in the United Kingdom.

If traditionalists question the historic legitimacy of the metric system, they need only consider Thomas Jefferson's efforts to devise a coherent system of measurements for the fledgling United States based on multiples of 10. The United States was already a leader on the coinage front in the English-speaking world, having abandoned the old pounds-shillings-pence system that survived elsewhere well into the 20th century in favor of the more manageable 100 cents to the dollar. Jefferson, one of the brightest intellects of his time, had been a U.S. diplomat in Paris when the French metric system took shape and had been impressed by its logic and coherence. As Secretary of State in 1790, when President Washington asked him to prepare alternative sets of measurement standards for consideration by Congress, Jefferson was ready. One of the two proposals—reportedly Jefferson's preference—was based on multiples of 10, just like the metric system, but Congress did not adopt it at that time.

Fahrenheit Meets Celsius

Metric today is not only the *lingua franca* of commerce, but of science as well. Pick up any medical journal, for instance, and all the quantities you see in the text will be in metric. Patients are described as being so-many centimeters tall, weighing so-many kilograms, and having a body temperature of so-many degrees Celsius (37 C is the same as "normal" body temperature of 98.6 F).

The relationship between Fahrenheit and Celsius is, in fact, the only conversion that most consumers may need to use that is somewhat tricky. But even it can be mastered, according to Gloria Marconi, a Silver Spring, Md., illustrator whose husband, Ercole, often asks her to make baked goods from recipes in the metric system cookbooks he brought with him from his native Italy. How to convert from Celsius to Fahrenheit for oven temperatures?

If your oven registers in Fahrenheit and the recipe tells you the baking temperature in Celsius, just multiply the Celsius number by 9 and then divide by 5. Add 32 to the result, and there's your Fahrenheit oven setting. (For example, if the cookbook calls for a Celsius setting of 220: 220 x 9 = 1,980 ÷ 5 = 396 + 32 = 428; 425 is close enough for your Fahrenheit oven setting.)

It's only human for Americans to be creatures of habit, but the inch-pound system habit has put us at a disadvantage with our competitors overseas and so has become a luxury we can no longer afford. The faster and more enthusiastically we embrace metrication, the more prosperous a nation we are likely to be. The Food and Drug Administration, with its product labeling requirements, is committed to do its part in this effort. ■

Questions

1. How many nations, in addition to the United States, do not conduct commercial affairs in metric units?

2. What is the advantage of "dual labeling" on packaging?

3. Why is the metric system easy to learn?

Answers begin on page 161.

3 *Discoveries come about through the observation of nature, by experimentation, or sometimes just by accident (serendipity). In the history of chemistry, there are numerous examples of serendipity from which many important developments have flowed. Serendipity becomes significant only when the observer recognizes that there is something that needs further exploration. A classic example was the accidental preparation of the mauve dye by W.H. Perkin in 1856. His discovery led to the beginning of the synthetic dye industry. Serendipity has played a key role in several new and varied applications of a remarkable element: selenium. From biological uses as an essential nutrient to industrial products like the new Verde Digital Film produced by the Xerox Corporation, selenium has a fascinating versatility.*

Serendipity

J.E. Oldfield

The development of uses for selenium has been marked by innovative conceptions, misinterpretations, and just plain accidents.

It was serendipity from the beginning for selenium, which was discovered by accident by Jons Jakob Berzelius in 1817. Berzelius owned an interest in a sulfuric acid manufacturing plant in Gripsholm, Sweden, and he became concerned that workers there appeared to be suffering from some work-related illness. After reviewing the processing methods, he determined that the problem had appeared after a change in the source of sulfur ore from an imported source to a local source that was more economical. Berzelius assumed that some impurity in the new ore might be responsible for the illness. He was looking for arsenic when he discovered a new element, selenium. The discovery was accidental but not immediately serendipitous because Berzelius did not find a ready use for this new element, which remained on laboratory shelves as a chemical oddity for a number of years.

More than most other elements, selenium has two distinctly different types of uses: in a number of varied industrial applications and as an essential nutrient. Let me begin with its industrial uses. Around 1873, more than 50 years after its discovery, selenium was found to be photosensitive; this finding led to its first practical application as a component of the selenium photocell. Although a number of applications have been found for these cells over the years, they have been overshadowed by the advent of xerography, wherein images are reproduced by light exposure of an electrostatically sensitized film of amorphous selenium. Not surprisingly, one of the world's largest users of selenium is the Xerox Corporation.

It is not recorded whether selenium's application in xerography was serendipitous—it may have been—but one of its most recent developments was. In November 1993, Xerox launched a new product called Verde Digital Film. During its production, the new material, which consisted of a polyester base coated with selenium, was charged and imaged and then accidentally overheated. To the surprise of the developers, this overheating caused the images to become permanent, and Verde Digital Film came into being. It involves a dry-process, silverless digital intermediate with exceptionally high resolution and produces high-quality printed documents faster and more economically than does conventional silver halide film. It will be used in mastering plates for offset printing and in the production of magazines, newspapers, and books.

When we became involved in investigating selenium's biological applications some 40 years ago, a friend suggested to me facetiously that, if selenium was so effective in reproducing images on paper, maybe it could similarly stimulate the reproductive processes in animals. We laughed, but the statement was prophetic: Selenium did prove to be essential for animal reproduction, as I will explain in detail later.

Reprinted with permission from *CHEMTECH*, vol. 25, no. 3, March 1995, pp. 52–55.
Copyright 1995 American Chemical Society.

In 1925 the first effective selenium rectifiers were developed in Germany (1) These were widely used until the development of semiconducting diodes, such as the silicon diode, in 1948. Before that time it had been thought that light interacted only with ionic crystals, not with solids such as selenium, and thus this use was unexpected. Research on selenium's conducting properties is ongoing. One topic of particular interest involves trigonal selenium, which is an arrangement of spiral chains in hexagonal symmetry (2).

Colors Appearing and Disappearing
In 1891 and 1892 Welz was granted German (3) and U.S. (4) patents, respectively, for the production of a brilliant rose-colored glass created using selenium, and soon a number of variations were demonstrated.

The many beautiful colors imparted to glass by selenium in various concentrations and combinations did not occur accidentally—they were the result of deliberate investigations—but the results were often unexpected. When accompanied by cadmium sulfide in the presence of a reducing agent, selenium produces a deep ruby color in zinc-potash glasses. In the presence of an oxidizing agent, however, selenium gives an amber tint to lead glass or a pinkish color to lime glass, whereas lime glass can be colored orange using selenium with cadmium sulfide plus a reducing agent (5).

Another serendipitous discovery was made in the development of a very different use for selenium in glass manufacture. Selenium-containing rose pigments were being used to mask the slight greenish tints caused by impurities of iron and copper in commercial glass. One day an excess amount of selenium was added, accidentally, to the melt. Surprisingly, all color disappeared, demonstrating that selenium, in appropriate quantities, could act as a glass decolorant (6). The food industry popularized the use of colorless glass to better show the natural colors of their products. Manganese had been used as a decolorizing agent, but selenium was found to hold some technological advantages over and cost less than manganese.

It was originally thought that the selenium decolorizing process was a purely physical reaction, but research has shown that the color of glass when both selenium and iron are present is different from glass in which either element is present alone (7). The decolorizing is apparently due to the formation of iron selenide, which effectively prevents the iron from imparting its usual greenish tint.

Glassmakers had to learn new rules for selenium processing. Whereas colors imparted by Co^{2+}, Cu^{2+}, and Ni^{2+} can be diluted to very faint tints, the colors selenium produces cannot be made paler by reducing the concentrations of the inherent cadmium, selenium, and sulfur (2). Weckerle (8) showed that by increasing the content of zinc oxide, selenium retention in the melt can be increased, resulting in a more brilliant red color. However, when the level of zinc oxide is raised above 18% (wt/wt), the equilibrium among some of the reactive sulfides is upset and the glass again becomes colorless. The applications of selenium in glass have expanded in recent years to include uses in modern building architecture and in so-called privacy glasses in vans and limousines. Gray and bronze heat-absorbing glasses contain 13 and 17 ppm of selenium, respectively (9).

Modifying Metals
Although selenium is classified as a nonmetal in generally accepted metallurgical terminology, useful applications have been found for it in metallurgy. The addition of selenium to steel improves its impact strength through interaction with the sulfides present (10) and also improves machinability (11).

In producing stainless steel, sulfur continues to be the preferred additive, although selenium has been used in some cases when corrosion resistance, hot and cold formability, and surface finish are important considerations. Selenium is also useful in processing other metals, such as copper, in which it improves machinability more than either sulfur or tellurium. It tends to impair weldability, although it appears not to interfere with either brazing or soldering qualities (12). The latest metallurgical application for selenium has been as a replacement for lead in producing plumbing brasses; this is an important advance in alleviating concerns about the health effects of lead leaching into drinking water (13). It is of interest that selenium, a known toxic agent, can be beneficial in preventing toxicity from other substances.

A Puzzle
These few examples of the many innovative uses of selenium in industry are interesting because of the diverse properties they illustrate. Little about selenium's colorant properties, for example, would suggest that it might serve as a semiconductor or improve machinability in the fabrication of metals. Additionally, selenium has proven during the past half-century to be just as versatile and puzzling in its biological applications.

Berzelius found, by implication, that selenium was toxic to humans when he identified it as a contaminant in the red sludge that formed during the synthesis of sulfuric acid. About 25 years later, selenium was documented to be toxic to animal life (14), and this finding was confirmed in the range country of the western United States almost a century afterward (15).

As with selenium's original discovery, serendipity characterized the discovery of its toxicity to livestock. Range animal poisonings, generally involving horses, had been called locally "alkali disease" on the assumption that they were caused by the animals drinking from alkali ponds in the area. Further investigation revealed, however, that the pond water was not involved at all; the poisoning resulted from animals eating certain types of range plants such as *Astragalus* (milk vetch) or *Machaeranthera* (woody aster), which accumulated selenium from the soil in toxic quantities.

Healthier Animals
Later, in the mid-20th century, the biological implications for selenium changed dramatically as a result of investigations by Klaus Schwarz, who was working at the National Institutes of Health. Schwarz had prepared concentrates of a substance that prevented liver necrosis in his laboratory rat colony and was attempting to extract the active principle from them. At that time, knowledge of nutrition was being rewritten through the discovery of what we now know as vitamins, and Schwarz was actually looking for a vitamin. By chance he identified selenium when, in the course of his investigations, his chief, De Witt Stetten, strolled into his laboratory, sniffed the air, and remarked that he detected a garlic-like odor that might be caused by selenium (*16*).

Even after Schwarz had noted selenium's involvement, he suspected that it might be functioning in organic combination, as part of a vitamin molecule. He had a number of possible organoselenium compounds prepared and tested them systematically against known selenium-responsive diseases. He sent us one—a valeric acid derivative—to test with sheep on diets that made them susceptible to "white muscle disease." We included it among several other treatments and found it was less effective than sodium selenite, a simple inorganic source of selenium, which I think disappointed him.

White muscle disease was a serious problem for livestock producers in Oregon's central high desert plateau when I came to Oregon State University in 1949. The condition, which takes its name from the bleached-out color in certain affected muscles caused by loss of myoglobin and deposition of calcium salts, annually resulted in the deaths of many newborn calves and lambs. When we began a systematic investigation of the problem, aware of Schwarz's recent research on somewhat similar symptoms, we included selenium among several different experimental treatments. We expected that the problem was due to a dietary deficiency of vitamin E and were surprised to find that selenium improved the condition and vitamin E did not (*17*): Serendipity again!

Commercial livestock producers were quick to recognize the cost-effectiveness of selenium supplements for their animals, because selenium not only prevented deaths attributed to white muscle disease but also stimulated weight gain and reproductive performance. U.S. Food and Drug Administration (FDA) officials initially prevented producers from applying the research results because they suspected that selenium might be carcinogenic; if so, its use in foods or feeds would have been (and would be) illegal under provisions of the Delaney amendment to the Food, Drug, and Cosmetics Act. Extensive research showed that selenium was not a carcinogen, and in 1974 the FDA authorized its use at a level of 0.1 ppm in the dry diet. The allowable level was increased to 0.3 ppm in 1987.

Growth of food animals raised in areas where soil, and therefore crops, are low in selenium content can be stimulated with carefully controlled dietary supplements (*18*). To some extent this discovery, too, was serendipitous because the growth increment first was noticed in records of animals that were primarily involved in investigations of myopathy (white muscle disease). Selenium is also an essential nutrient in animal reproduction: It reduces embryonic deaths in females (*19*) and aids sperm motility in males (*20*). Its enhancement of reproduction in sheep allowed the development of unimproved hill lands in New Zealand, where sheep production previously had been uneconomical. Dietary selenium supplements have become an accepted livestock production practice worldwide.

Healthier People
The selenium story took another unexpected twist with the report of evidence showing that mortality in certain types of cancer was greater in areas where soil and food selenium levels were very low (*22*). Instead of being a carcinogen, selenium was a protective agent. This preliminary suggestion, dealing as it did with one of the most dreaded human diseases, stimulated a tremendous surge of research, the results from which were largely confirmatory (*23*). This effort is ongoing, and last year a large and well-controlled study in an area of high cancer incidence in China (*24*) included selenium with some other antioxidants in a successful treatment regimen against cancer of the digestive tract. Selenium's protective effect against cancer is now believed to be so promising that the government is funding research to develop new organoselenium compounds for cancer prevention (*25*).

The source materials for the human diet in most industrialized countries are sufficiently varied that the occurrence of selenium deficiency is unlikely; however, there are some exceptions. In parts of China localized populations grow most of their own food, and in such areas with selenium-deficient soil there has been a high incidence of a cardiomyopathy called Keshan disease, which is selenium responsive. The discovery of the treatment for this condition is serendipitous, given that it originally was developed for use with animals. When I visited China in 1978 I attended a seminar at the Chinese Academy of Medical Sciences in Beijing, where I learned about their successful control of Keshan disease. When I asked how they had determined the effective dosage level of selenium, they responded, "We used the same level you used in your studies with sheep."

Two smaller countries, Finland and New Zealand, have extensive areas of selenium-deficient soil, and their governments are sufficiently concerned about the problem of selenium deficiency in animals and humans that they have either mandated or encouraged the addition of selenium to fertilizers used in the production of food and forage crops. This situation has been carefully monitored (26) and the practice of supplementing soil with selenium has been both nutritionally effective and environmentally responsible.

A Continuing Story

One fascination of the selenium story is its continuity; there seems to be no end to the new and different applications for this remarkable element. In its biological use, past successes tend to make serendipity's recurrence less likely than in the past. More frequently today, selenium is included in experimental protocols deliberately rather than by accident. A recent example is the investigation of possible effects of selenium in the immune process that protects against animal and human disease (27), which seems poised to usher in a whole new array of uses for selenium, apparently at somewhat higher dietary levels than those conventionally required for nutrient function.

Certainly much of selenium's usefulness, both biological and industrial, serves as a testament to the powers of serendipity and argues strongly for scientists to avoid becoming too narrowly focused and to apply imagination and innovation in their research.

References
(1) Presser, E. *Funkbastler* **1925,** *44,* 558.
(2) Stuke, J. In *Selenium;* Zingaro, R.A.; Cooper, W.C., Eds; Van Nostrand Reinhold: New York, 1974; p. 174.
(3) Welz, F. German Patent 63 558, 1891.
(4) Welz, F. U.S. Patent 479 689, 1892.
(5) Silverman, A. *J. Am. Ceram. Soc.* **1928,** *11,* 81.
(6) Hirsch, M. U.S. Patent 576 312, 1897.
(7) Gooding, E.J.; Murgatroyd, J. *J. Soc. Glass Technol.* **1935,** *19,* 43.
(8) Weckerle, H. *Glastech. Ber.* **1933,** *11,* 273.
(9) La Course, W.C. In *Proceedings of the 5th International Symposium on Uses of Se and Te, Brussels, Belgium;* Selenium/Tellurium Development Association: Grimbergen, Belgium, 1994; p. 263.
(10) Almand, E. *Rev. Metall. Paris* **1969,** *66,* 749.
(11) Carapella, S.C., Jr.; Aborn, R.H.; Cornwell, L.R. In *Selenium;* Zingaro, R.A.; Cooper, W.C., Eds.; Van Nostrand Reinhold: New York, 1979; p. 762.
(12) *Machining of Copper and Its Alloys;* Copper Development Association: London, 1961; publication 34.
(13) King, M.G.; Hayduk, D.; Mirza, A. *Proceedings of the 5th International Symposium on Uses of Se and Te, Brussels, Belgium;* Selenium/Tellurium Development Association: Grimbergen, Belgium, 1994; p. 51.
(14) Japha, A. Ph.D. Dissertation; Halle, Germany, 1842.
(15) Moxon, A.L. "Alkali Disease"; South Dakota Agr. Exp. Sta.: Brookings, SD, 1937; bulletin 311.
(16) Jukes, T.H. *J. Appl. Biochem.* **1983,** *5,* 283.
(17) Muth, O.H.; Oldfield, J.E.; Remmert, LF.; Schubert, J.R. *Science* **1958,** *128,* 1090.
(18) Oldfield, J.E.; Muth, O.H.; Schubert, J.R. *Proc. Soc. Exp. Biol. Med.* **1960,** *103,* 799.
(19) Hartley, W.J.; Grant, A.B. *Fed. Proc.* **1961,** *20,* 679.
(20) Wu, A.S.H.; Oldfield, J.E.; Whanger, P.D.; Weswig, P.H. *Biol. Reprod.* **1973,** *8,* 625.
(21) Paulsson. K.; Lundbergh, K. *Proceedings of the 5th International Symposium on Uses of Se and Te, Brussels, Belgium;* Selenium/Tellurium Development Association: Grimbergen, Belgium, 1994; p. 287.
(22) Shamberger, R.J.; Frost, D.V. *Can. Med. Assoc. J.* **1969,** *100,* 682.
(23) Mertz, W. In *Selenium in Biology and Medicine;* Van Nostrand Reinhold: New York, 1984; Vol. A, pp. 3–8.
(24) Blot, W.J.; Li, J.Y.; Taylor, P.R.; Gus, W.; Dawsey, S.; Wang, G.L.; Yang, C.S.; Zheng, S.F.; Gail, M.; Li, G.Y.; Yu, Y.; Liu, B.; Tangrea, J.; Sun, Y.H.; Liu, F.; Fraumeni, J.F., Jr.; Zhang, Y.H.; Li, B. *J. Natl. Cancer Inst.* **1993,** *85,* 1483.
(25) Reddy, B.S.; Revenson, A.; Kulkami, N.; Upadhyaya, P.; El-Bayoumy, K. *Cancer Res.* **1992,** *52,* 5635.
(26) Kivisaari, S.; Vermeulen, S. In *Proceedings of the 5th International Symposium on Uses of Se and Te, Brussels, Belgium;* Selenium/Tellurium Development Association: Grimbergen, Belgium, 1994; pg 95.
(27) Spallholz, J.E.; Boylan, L.M.; Larsen, H.B. *Ann. NY Acad. Sci.* **1990,** *587,* 123. ■

Questions

1. What results from the presence of selenium in zinc-potash glass?

2. How does the addition of selenium to steel improve the alloy?

3. How do selenium supplements benefit livestock?

Answers begin on page 161.

4 *Many Americans may be consuming too little calcium and increasing their risk of osteoporosis. This loss of bone mineral density leads to spongy, brittle bones that break more easily than normal bones. Most calcium, which is the most abundant mineral in the body, is deposited in the bones, where it provides strength and durability. A small amount is also found in the blood and cellular fluids, helping to regulate many important functions, such as blood clotting, nerve impulse transmissions, and muscle contraction. The calcium in your diet eventually helps to control these processes. A better understanding of calcium and adequate intake can help you avoid serious health consequences.*

Calcium: Here's How to Bone Up on This Essential Mineral

What do you and a 10-year-old have in common?

You both need about the same amount of dietary calcium.

Calcium's role in bone health isn't limited to growing children and teens. In adults, eating too little calcium is linked with osteoporosis. Each year, more than 1.5 million Americans have fractures related to thin, brittle bones.

Last June, scientists convened by the National Institutes of Health (NIH) examined new information about calcium's role in preventing osteoporosis. Their consensus: The recommended dietary allowance (RDA) doesn't adjust for your calcium needs as you grow older.

How Much Is Enough?
The amount of calcium you need varies throughout your life. Greatest needs occur during the period of rapid growth among children and adolescents and among pregnant and nursing women.

After about age 30, the amount of bone you form typically reaches its maximum. Bone formation and bone loss are balanced. For this reason, the RDA for calcium drops from a high of 1,200 milligrams for adolescents to 800 milligrams for adults older than 25.

However, in light of evolving research into calcium-related disorders, the NIH panel recommended higher calcium intakes for adults. (They also recommended more calcium for adolescents to reduce their risk of osteoporosis as they grew older.)

The panel agreed you need more calcium because:
• *You absorb less calcium*—Calcium absorption decreases as you age, especially after about age 65. You also make less vitamin D—essential for enhancing the amount of calcium that ultimately reaches your bones (see *Mayo Clinic Health Letter,* December 1994).

If you're a woman:
• *Your estrogen level falls*—Estrogen slows calcium loss from bones. At menopause when your estrogen level drops, bone loss accelerates.

During the first six to eight years of menopause, estrogen replacement therapy slows bone loss. That's why women who take estrogen need less calcium than women who don't take the hormone (see "Consensus: You need more calcium").

After about 10 years, estrogen's effects are less dominant and calcium's effects increase. Supplemental amounts of calcium in the range of 1,500 milligrams seem to reduce bone loss.

Ways to Get More Calcium
To meet the higher calcium recommendations:
• *Choose foods first*—Dairy products are your richest sources of calcium (see "15 best calcium sources"). Select low-fat items such as skim milk

---

Consensus: You Need More Calcium

A National Institutes of Health panel recommends these daily calcium levels (in milligrams) as protection against osteoporosis:

Women
25 to 50	1,000
50 to 65	1,500
On estrogen	1,000
65 plus	1,500

Men
25 to 65	1,000
65 plus	1,500

Reprinted from the May 1995 *Mayo Clinic Health Letter* with permission of Mayo Foundation for Medical Education and Research, Rochester, Minnesota 55905. *Mayo Clinic Health Letter* does not specifically endorse any company or product.

or low-fat yogurt to limit calories and fat.

If you don't or can't drink milk, some kinds of leafy green vegetables and legumes, plus calcium-fortified products, are other ways to boost your calcium intake.

To fortify your own foods, add a tablespoon or two of nonfat dry milk to baked goods, casseroles, meatloaf or hot beverages.

• *Consider a supplement*—Depending on your diet, food alone can provide you with recommended amounts of calcium. If your diet doesn't include dairy products, however, you may need a calcium supplement (see "How to select a calcium supplement").

Tips for Taking a Supplement
Food is the best way to get calcium because it contains a variety of essential nutrients. But if you need a supplement:
• *Take small doses*—Limit single doses to no more than 600 milligrams of elemental (available) calcium. Your body absorbs small doses best.
• *Take with meals*—Although some foods may interfere with calcium absorption, taking a supplement with meals is most convenient. Many older adults also have reduced levels of stomach acid. By stimulating acid production, eating enhances calcium absorption.
• *Add vitamin D*—If you're not taking a multivitamin, choose a calcium supplement that also provides 200 to 400 International Units (IU) of vitamin D.

Keep Calcium in Perspective
Getting enough calcium in your diet may help slow bone loss and reduce your risk of osteoporosis. But remember, regular weight-bearing exercise also helps keep your bones strong. And if you're a woman, estrogen replacement, combined with exercise and adequate dietary calcium, offers the best defense against bone loss and fractures.

---

15 Best Calcium Sources
Along with three cups of milk, have a serving or two of any of these foods to get more than 1,000 milligrams of calcium.

Food	Calcium (milligrams)
Milk (skim and low-fat), 1 cup	300
Tofu set with calcium, 1/2 cup	258
Yogurt, 1 cup (average of low-fat brands)	250
Orange juice (calcium-fortified), 8 ounces	240
Ready-to-eat cereal (calcium-fortified), 1 cup	200
Mozzarella cheese (part-skim), 1 ounce	183
Canned salmon with bones, 3 ounces	181
Collards, 1/2 cup cooked	179
Ricotta cheese (part-skim), 1/4 cup	169
Bread (calcium-fortified), 2 slices	160
Cottage cheese (1 percent fat), 1 cup	138
Parmesan cheese, 2 tablespoons	138
Navy beans, 1 cup cooked	128
Turnips, 1/2 cup cooked	125
Broccoli, 1 cup cooked	94

How to Select a Calcium Supplement
To improve absorption, calcium supplements are available as various salts such as carbonate, citrate and phosphate. You'll see most products as calcium carbonate, but all preparations are well-absorbed. With the exception of Tums, the following products are available in varieties that also provide vitamin D.

We don't recommend calcium supplements made from bone meal or dolomite because they may be contaminated with lead, mercury or arsenic.

Supplement	Calcium/pill (milligrams)	Advantages
Calcium carbonate (Caltrate, OsCal, Spring Valley, Tums)	250, 500 or 600	Least expensive and most available
Calcium citrate (Citracal)	200 or 315	Easily absorbed regardless of the amount of stomach acid
Calcium phosphate (Posture)	600	Least likely to cause constipation ■

Questions

1. Compare an adult's need for dietary calcium to a child's need.

2. Why should older adults take in more calcium than younger adults?

3. Why are women especially susceptible to bone loss after menopause?

Answers begin on page 161.

5 *Platinum may be more expensive, but gold remains the noblest metal in the eyes of chemists. Elements like those in group 18, which tend to remain in an elemental state, resisting reaction with other substances to form compounds, are said to be noble. The other metals in group 11(1B) react fairly easily with their environment—silver items tarnish, and a copper roof turns green—but gold maintains its beautiful luster. The ability of gold-covered objects to resist surface changes or tarnishing is part of the reason it has fascinated kings and commoners for centuries. The gold mask of Tutankhamen is still a brilliant yellow, just as it was 3,500 years ago at the pharaoh's death.*

Gold: Noblest of the Nobles

Ken Reese

Imagine, if you will, a prospective investor being told that "Gold, atomic number 79, is a third-row transition metal in Group IB of the periodic table..."

The properties of gold made it apparently the first metal to be used by man and certainly the cause of much skulduggery ever since. Archaeologists found the earliest gold jewelry known—dating from about 5,000 years ago—in the royal Sumerian tombs at Ur in what is now southern Iraq (*1*). The death mask of Tutankhamen, the boy pharaoh of Egypt, shows that skilled goldsmiths were at work there at least 3,500 years ago. The Chinese and Egyptians were using gold in dentistry at least 3,000 years ago. The search for gold, although the source of much grief, has done much to catalyze world exploration and the development of world trade. The centuries-long, if unavailing, efforts of alchemists to convert base metals to gold paved the way for the emergence of modern chemistry in the mid-seventeenth century. For millennia gold has served as money or as a guarantor of money. In recent years, besides its traditional uses in jewelry and other decorative functions, the metal has found important industrial applications, especially in electronics. And today, as always, people involved closely with gold tend to keep a weather eye out for the vagaries of politics.

A Rare Treasure
Imagine, if you will, a prospective investor being told that, "Gold, atomic number 79, is a third-row transition metal in Group IB of the periodic table. It occurs naturally as a single stable isotope of mass 197. Its electronic configuration is [Xe]$4f^{14}5d^{10}6s^1$. Common oxidation states are 0, 1, and 3" (*2*). Such data would rarely be powerful sales points, but they do explain some of the element's more engaging characteristics. Gold is the noblest of the noble metals; it resists corrosion forever in ordinary conditions and is found naturally in highly pure form. Gold also occurs as calaverite, or gold telluride (AuTe$_2$), first identified in Calaveras County, CA, home of Mark Twain's celebrated jumping frog. Gold does not react with tellurium, however, below about 475 °C, so the telluride found at the earth's surface must have formed in deeper, hotter zones. Gold is remarkably ductile: One gram can be drawn to 165 m of wire with a diameter of 0.02 mm, and gold leaf that is only 0.14 μm thick retains the metal's characteristic sheen and color.

People treasure gold not only because of its appearance, ductility, and indestructibility, but also because of its rarity. The earth's crust, on average, contains about 3.5 ppb of gold, and seawater averages on the order of 10 pptr. Larger, local concentrations of metallic gold occur in sedimentary and igneous rocks in combination with materials such as quartz, silver and base metal sulfides, selenides, and tellurides. The gold taken from the earth during recorded history totals only about 125,000 metric tons (a metric ton, or tonne, is 1,000 kg) (*3*). That amount would fit into the bottom third of the Washington

Reprinted with permission from *Today's Chemist at Work*, vol. 3, no. 10, October 1994, pp. 47–48. Copyright 1994 American Chemical Society.

Monument. With a specific gravity of 19.31, gold is 70 percent denser than lead and the seventh densest naturally occurring element. Very close to gold is tungsten with a specific gravity of 19.35; miscreants have been known to create ersatz gold bars by filling voids formed in the real material with tungsten topped off by gold plate.

Going for Gold
Weathering and erosion in flowing streams free gold from rocks as fine grains and nuggets, or placer gold. Discoveries of placer deposits triggered the gold stampedes of 1848–1900 in California, Australia, South Africa, and the Klondike. Miners collect placer gold by picking up nuggets or swirling goldbearing soil with water in a pan or sluice box until the small gold particles sink to the bottom. The source of most new gold, however, is rock containing from one to about thirty grams of the metal per ton. The recover the gold, the rock is ground to suitable fineness and then treated by one or more of several processes: amalgamation with mercury, cyanidation, concentration by gravity or flotation, and roasting. A breakthrough for low-grade ore came in 1887, when physicians Robert and William Forrest and chemist John MacArthur patented a process for extracting gold from South African ore. The finely ground ore is leached with a sodium cyanide solution to form $Na[Au(CN)_2]$. Zinc, added as dust, displaces the gold, which precipitates.

High gold prices spurred a new gold rush in the 1980s with a focus on low-grade ores (down to one gram per ton) and new technology. One new method, heap leaching, requires only that the ore be crushed into chunks, which is much cheaper than milling it to the fine powder required for the MacArthur–Forrest process. Sodium cyanide solution percolates for several weeks through crushed ore heaped on leach pads. In a second innovation, the gold-rich leach-pad solution is pumped over carbon granules (preferably from coconut shells), which adsorb the gold. The metal is stripped from the carbon with an alkaline cyanide solution and recovered electrolytically. The cyanide solution and carbon are recycled. By the late 1980s, heap leaching/carbon adsorption accounted for half the gold produced by US mines and was finding much use overseas as well. A third new technology is roasting to oxidize refractory sulfide ores so the gold can be recovered conventionally.

The leading gold producers in 1993 were South Africa (620 tonnes), the US (336 tonnes), Australia (247 tonnes), and the former Soviet Union, now the Commonwealth of Independent States (244 tonnes). Mine output met some 64 percent of world demand for gold last year; old gold scrap, mainly from jewelry and electronic devices, met about 15 percent; sales of existing stocks, public and private, accounted for the remaining 21 percent. Global demand totaled 3,538 tonnes. Of this amount, 2,989 tonnes, or 84 percent, were used in fabricated products: jewelry, electronic equipment, dental materials, other industrial products, and medals and coins.

For More Than Jewelry
Although electronics accounted for only about 6 percent of the gold used in fabricated products in 1993, the metal's ductility, corrosion resistance, and thermal and electrical conductivity make it the best material for certain solid-state applications. The main gold products are bonding wire, used as internal connections in semiconductor integrated circuits, and gold salts used to electroplate contacts for switches, relays, and various kinds of connectors. Gold's ductility permits bonding wire to be drawn to a diameter of 0.01 mm. Gold-plated contacts do not acquire the surface oxides that can impede the flow of the typically small currents in solid-state devices. Gold bonding wire and gold-plated contacts are found in equipment as diverse as personal computers, telephone jacks, home appliances, and automotive air bag actuators.

A Modern Gold Rush
The gold rush of recent years involved major changes in the geographical sources of mined gold. Although South Africa remains the top producer, its share of mine production in the West fell from 75 percent twenty years ago to about 30 percent last year. US output in 1993 increased for the fourteenth consecutive year, to a share of 18 percent; Australian, Canadian, and Latin American (total) outputs also have grown steadily during the past decade. The US balance of trade in gold rose from a $1.6 billion deficit in 1980 to a $1.9 billion surplus in 1993. Still, the geographical pattern of gold mining characteristically changes, and a new shift seems to be well under way.

Where money is spent on exploration suggests where new mines will be found. In this vein, seventeen producers surveyed by The Gold Institute (Washington, DC) spent 71 percent of their exploration dollars in this country in 1989 and only 54 percent in 1993; last year's exploration, moreover, was aimed mainly at expanding mines, not finding new properties. These producers, who account for about three-quarters of US production, see other countries becoming more hospitable to their operations.

One indicator of where exploration dollars will be spent is the impending revision of the federal Mining Law of 1872. The law permits qualified applicants to acquire title to public lands containing hard-rock minerals, such as gold, for $5 or less per acre and thereafter pay no royalties on the minerals extracted. Congressional efforts to modernize the act were highlighted last May, when a subsidiary of the Canadian-based American Barrick Resources Corp. acquired for less than $10,000 some 1,800 acres of public land in Nevada containing an estimated $10 billion in gold. Interior Secretary Bruce Babbitt, who turned over the title to the company, termed the transaction "the biggest gold heist since the days of Butch Cassidy." Babbit conceded that "these folks stole it fair and square" but did not mention the $1 billion that Barrick had spent exploring the area and developing the mine.

In June, House and Senate versions of legislation reforming the mining act went to conference.

Points to be resolved include the size of the royalty and the claim maintenance fee as well as environmental requirements. The industry agrees that the act should be reformed, according to John Lutley of The Gold Institute, but fears being overburdened with costs and environmental restrictions. "It happened in Canada about five years ago," he says, "and gold mining there shows clear signs of decline." As matters stand, Lutley believes, US gold output will peak in perhaps five years at about 370 tonnes annually and decline gradually during the subsequent fifteen years.

The hot new area, meanwhile, is Latin America. Beginning with Chile in 1982, nine Latin American nations have reformed their laws to attract investment in their mining industries. The area is geologically promising, and much of it is unexplored. The seventeen companies surveyed by The Gold Institute spent $58 million on exploration there in 1993, more than four times the amount spent in 1989.

References
1. Green, T. *The World of Gold*; Rosendale Press: London, 1993.
2. *Kirk–Othmer Encyclopedia of Chemical Technology*; John Wiley and Sons: New York, 1980; 3rd ed., Vol. III.
3. *Gold 1994*; Gold Fields Mineral Services Ltd.: London, 1994. ■

Questions

1. Why is gold considered the noblest of the metals?

2. What properties of gold make it especially useful in the electronics industry?

3. What home equipment or appliances might contain electronics using gold wire or contacts?

Answers begin on page 161.

6 *In 1994, researchers at GSI in Darmstadt, Germany, extended the periodic table to include the newly synthesized elements 110 and 111. In the 1980s, the same group of scientists prepared elements 107, 108, and 109. These transfermium elements—those elements above 100—are produced one atom at a time in accelerators by bombarding targets with energetic beams of ions. In their unique and latest approach, the GSI team uses an accelerator to fuse medium-weight elements (the beam) projected at other medium-weight elements (the target). All of the transfermium elements are unstable, decaying spontaneously. With half-lives of only a few minutes or, in some cases, only milliseconds, the difficulty of preparing and identifying these heavy elements is enormous. However, the GSI scientists are confident that their new technique will soon enable them to prepare element 112.*

Heavy-Ion Research Institute Explores Limits of Periodic Table

Michael Freemantle

German scientists are confident that element 112 will soon be created, followed eventually by elements 113 and 114.

German nuclear scientists are extending the periodic table to its limits. After creating elements 107, 108, and 109 in the early 1980s, the team at what is known as the GSI in Darmstadt synthesized elements 110 and 111 late last year. And the GSI researchers are confident that they will be able to add element 112 to the periodic table by early 1996.

The group at GSI, which stands for Gesellschaft für Schwerionenforschung (Society for Heavy-Ion Research), aims to make element 112 by fusing zinc and lead atoms. In the longer term, they also hope to synthesize elements 113, 114, and possibly 115.

"We are confident of eventually being able to produce the heavier elements, but it will be nice to know where the end is," says physicist Peter Armbruster, head of the nuclear chemistry department at GSI.

Armbruster explains that the syntheses of these elements are also providing valuable insights into nuclear structure. The superheavy element 114 is of particular interest. Nuclear scientists predict that the isotope of element 114 with atomic mass 298 should lie at the center of an "island of stability" and thus have a significantly longer half-life than the heaviest elements known today.

"We are trying to find the upper limit of stability of heavy nuclei," says physicist Sigurd Hofmann, a group leader at GSI. "This will lead to a better understanding of the creation of heavy elements in the universe."

Elements 101 onward are known as the transfermium elements. They are synthesized by the fusion of the nuclei of two lighter elements. The compound nuclei that are formed are very unstable and therefore exist only fleetingly.

Elements 101 and 106 were discovered between 1955 and 1974 by groups at Lawrence Berkeley Laboratory and Lawrence Livermore National Laboratory, both in California, and the Joint Institute for Nuclear Research, Dubna, Russia. The groups produced these heavy elements by fusing the nuclei of slightly less heavy elements, such as the actinides californium (element 98) and einsteinium (element 99), with light particles such as carbon and oxygen ions. These light particles were projects with high energies at the actinide targets.

The approach used by GSI scientists to create elements 107 to 111 is different from that used by the U.S. and Russian laboratories for elements 101 to 106. GSI projected medium-weight elements with lower energies at other medium-weight elements.

The GSI approach, which was first developed by the Dubna group and tested on several fermium (element 100) and transfermium isotopes, is sometimes called "cold fusion." This term is used because the excitation energies of the compound nuclei are lower than those of the same compound nuclei produced in the bombardment of targets with much higher atomic numbers by lighter ions.

"The process of fusion is a two-step process," explains Hofmann. "Fusion of a projectile nucleus with a target nucleus results in a compound nucleus with an excitation energy, or temperature. It then cools down to arrive at the ground state by emitting particles, usually neutrons."

The cold-fusion approach reduces the probability of the nucleus undergoing spontaneous fission. By careful control of experimental conditions, the GSI group "can ensure that excitation energies are very small so the only one neutron is emitted from the nucleus," according to Hofmann.

The GSI group does not use the term cold fusion, however, because it also is used for the highly controversial fusion of deuterium nuclei reported in March 1989.

Armbruster and Hofmann led the team that synthesized elements 110 and 111 (C&EN, Nov. 28, 1994, page 5, and Jan. 2, page 7). The team consisted of six GSI researchers and six guest scientists, from the Dubna laboratory; the department of nuclear physics, Comenius University, Bratislava, Slovakia; and the department of physics, University of Jyväskylä, Finland.

The group produced the first atom of element 110 on Nov. 9, 1994. In a two-week experiment, the team bombarded a lead-208 target with a beam of nickel-62 atoms. Four atoms of ^{269}110 were created. In a separate experiment, during the following weeks, the team also created nine atoms of the ^{271}110 isotope by bombarding the lead-208 target with atoms of the neutron-rich nickel-64 isotope.

On Dec. 8, 1994, the same team created the first atom of element 111 by using nickel-64 projectiles and a bismuth-209 target. In all, three atoms were synthesized during the 18-day experiment. The atoms have a half-life of around 1.5 milliseconds and a mass of 272.

"The creation of the heavier isotope of 110 with nickel-64 was a very positive result for us," says Hofmann. "The cross section was four times higher than that with nickel-62, although we do not completely understand this increase."

Cross section is a measure of the probability that a nuclear reaction will take place. It is equivalent to the surface area presented by the atomic nucleus of a target to an incoming particle. The unit of cross section is the barn, which equals 10^{-24} sq cm.

"We see a logarithmic-scale linear decrease of the cross sections going from element 102 up to element 109," says Hofmann. The cross sections for production of elements 110 and 111 are about 3 picobarns. One picobarn is 10^{-12} barns or, as Armbruster points out, the probability of hitting 1 sq mm in an area of 1 sq km.

"With our beam intensities and efficiencies, 1 picobarn means that we produce on average one atom in six days," explains Hofmann. "But to be on the safe side, we use a beam time that is three times longer."

Cross sections for many nuclear reactions are highly dependent upon the energy of the incident particle. There is a very narrow band of energies on either side of which the probability of fusion rapidly decreases. The projectiles have to be accelerated to the precise velocity needed to just overcome the fusion or Coulomb barrier of repulsive electrostatic forces between the two colliding nuclei. If the energy is too low, there is no fusion; if it is too high, fission of the fused nucleus occurs.

"We try to find the maximum production cross section as a function of beam energy," says Hofmann. "We determined the whole distribution of production probabilities as a function of beam energy for elements 104 and 108. With these two values, we were able to extrapolate to the element 110."

The GSI researchers were thus able to calculate the optimum beam energy required for the nickel-62 projectiles used to create ^{269}110. "We didn't know the best projectile energy for the nickel-64 beam," says Hofmann. "We therefore began with nickel-62 and only later changed to nickel-64."

Armbruster observes that "The high production rate for element 110 with nickel-64 gives us new hope that we can make element 112 and eventually elements 113 and 114."

The GSI team of researchers that created elements 110 and 111 are now planning to synthesize element 112 by bombarding a lead-208 target with beams of zinc-68 and zinc-70 projectiles in separate experiments. If the synthesis is successful and "the cross section does not decrease further by extrapolation, then we are hopeful that we can also make element 114 by bombarding lead with germanium-74 or -76," says Hofmann. In the long term, the team may well attempt to fuse zinc-70 and germanium-76 isotopes with a bismuth-209 target to form elements 113 and 115, respectively.

The cross sections for producing these heavy elements are approaching the limits of experimental sensitivity. Beam times of several months would have been necessary to synthesize a few atoms of elements 110 and 111 with the experimental setup at GSI used in the early 1980s to create elements 107 and 109. "Since 1988, we have made continuous improvements so that our experiments are now at least 10 times faster than the old experiments carried out between 1980 and 1985," says Hofmann. "For example, in 1984 we made three atoms of element 108. When we recently repeated the experiment, we found a total of 75 atoms in a comparable measuring time."

Before commencing the experiment to make element 112, the group needs to make further improvements to the experimental setup in the area close to the target, which will take until the end of this year. "I think we can do the 112 experiment at the beginning of next year," Hofmann says.

The team now uses a so-called electron cyclotron resonance (ECR) source to generate the ions for the beam. "This ion source has improved stability and uses less material," says Hofmann. "This is important because we use enriched isotopes as projectiles and these are quite expensive. With the ECR source, we only need a few milligrams per hour."

The nickel-64 beam, for example, is generated from metallic nickel enriched to 93.1%. Projectile $^{64}Ni^{9+}$ ions extracted from the ECR source are accelerated to the energies needed for reactions at the Coulomb barrier. GSI's 120-meter-long high-frequency heavy-ion accelerator—known as UNILAC—is used to accelerate the ions. With this accelerator, it is possible to continually adjust the velocity of the ions. The average beam current is 3×10^{12} projectiles per second.

To synthesize element 111, eight bismuth-209 targets were mounted on a wheel rotating at 1,125 rpm. The targets are made by evaporating bismuth, chemically purified to 99.99%, onto a carbon backing.

The Darmstadt group has developed a technique that allows the recoil velocity imparted to the reaction products by the incoming projectiles to transport the newly formed nuclei to the detector. A nucleus created by fusion of a projectile ion and a target atom continues along the projectile ion's original path at a velocity up to several percent of the speed of light. This means that nuclei with half-lives as low as 100 nanoseconds can be detected.

Most projectiles striking the target go straight through without reaction and are absorbed by a copper plate beam stop. Atoms of a new element are heavier than the projectiles and therefore have a lower velocity. These slower in-flight fusion products are separated from the projectiles by electric and magnetic fields in an 11-meter-long velocity filter known as SHIP (separator for heavy-ion reaction products), built at GSI in the 1970s in collaboration with the physics department of the University of Giessen, Germany.

"Over the past five years, we have improved the background suppression of SHIP," says Hofmann, who leads the SHIP group at GSI. "The separation of reaction products by the velocity filter is now highly efficient."

The fusion products are guided to a sophisticated detector system consisting of two time-of-flight detectors, an array of 16 position-sensitive silicon detectors, and germanium detectors. Three simultaneous signals, one from each of the time-of-flight detectors and one from the silicon detectors, indicate implantation of an ion. Some implanted ions are excited after α-decay and emit γ-rays. These rays are detected by the germanium detectors.

With the signals from this detector system, the team can correlate implanted reaction products with their subsequent radioactive decay chains. Each fusion product has its own fingerprint characterized by the lifetimes of the isotopes and the energies of the particles emitted in the decay chain.

Elements 110 and 111 both decay by emitting a succession of α-particles (^{4}He nuclei). The GSI scientists were able to follow the decay chain of element 110 down to nobelium (element 102) and that of element 111 down to lawrencium (element 103). "Long decay chains are needed to characterize an implanted reaction product. Often one α-decay is not enough," says Hofmann.

Armbruster attributes the successful synthesis of these two elements to the improved sensitivity of the GSI equipment. "In 1994, we started with equipment that was 12 times more sensitive than we had before," he says.

Even greater sensitivity is required for elements 112 onward, as these have relatively low cross sections. GSI extrapolations show that the reactions of zinc-68 and zinc-70 with lead-208 to form element 112 have cross sections of 1 and 4 picobarns, respectively. For element 114, the cross sections reduce to 0.2 and 0.1 picobarn with germanium-74 and -76, respectively, as projectiles.

Element 114 is a key target in this work. Nuclear scientists predict that it is one of a series of elements with closed-shell nuclear structures. These structures have "magic numbers" of protons and neutrons that fully occupy the shells. Lead is one of the elements in this series. The isotope lead-208, which is very stable, has 82 protons and 126 neutrons. Both are magic numbers.

Protons and neutrons in closed shells have high binding energies. The nuclei are therefore particularly stable in much the same way that the noble gas elements, with their full shells of electrons, are stable. The next closed shells after lead occur at 114 protons and 184 neutrons.

Hofmann points out that element 114 isotopes with more than 170 neutrons all lie in the island of stability. All should have relatively long half-lives. Fusion of germanium-76 with lead-208 should produce an isotope with 170 neutrons.

"We want to make these isotopes to test our ideas on nuclear structure," say Armbruster. "Elements above 106 in the periodic table exist solely because of shell-stabilizing effects of the additional binding energy arising from internal quantum mechanical ordering of the protons and neutrons in the nuclei."

He explains that nuclei can have either stable spherical or less stable ellipsoidal configurations. "The nucleus of element 114 should be spherical, whereas that of element 110 is deformed, that is, ellipsoidal," Armbruster says.

The GSI researchers have not yet proposed names for elements 110 and 111, and it is not absolutely certain that they were the first to synthesize element 110. There is some evidence that one atom of element 110 with mass 267 may have been created at Lawrence Berkeley Laboratory in 1991 by bombarding bismuth with cobalt projectiles [*Nature,* **373,** 471 (1995)]. And in 1994, a joint U.S.-Russian team conducted experiments at the Dubna facility to synthesize element 110 by "hot fusion" of sulfur-34 projectiles with plutonium-244 target atoms. According to the *Nature* report, the researchers have "identified one decay chain atttributable to ^{273}110, but are still analyzing the data."

"The problem is not only synthesis of an element but also identification," comments Armbruster. "Compared with our experiments, the other experiments are, for us, not at all convincing—but this conclusion should be reached by an independent commission."

Although there can be no doubt that the GSI group discovered elements 107 to 109, the researchers are not happy that the International Union of Pure & Applied Chemistry (IUPAC) made changes to their proposed names for these elements. In 1992, the group proposed nielsbohrium for 107, hassium for 108, and meitnerium for 109. IUPAC decided on bohrium for 107, hahnium for 108, and meitnerium for 109. The GSI group uses the name hahnium for element 105.

"IUPAC should not change our proposed names without asking us," says Armbruster. "We should not read in the newspapers about IUPAC's decisions on the names of the elements we have discovered."

The synthesis of new elements and the names eventually assigned those elements generate newspaper stories. However, the activities of the GSI nuclear chemistry department extend beyond syntheses of isotopes of the transfermium elements.

A group in the department, led by chemist Matthias Schädel, is participating in a series of collaborative experiments investigating the chemistry of the transactinide elements—elements 104 and onward. The other teams involved are led by Jens V. Kratz at the Institute for Nuclear Chemistry, University of Mainz, Germany; Heinz Gaggler at Paul Scherrer Institute, Villigen, Switzerland; and Darleane C. Hoffman at Lawrence Berkeley Laboratory, University of California, Berkeley (C&EN, May 2, 1994, page 24).

The group studying transactinide chemistry "is currently investigating the chemical properties of element 105 to see how it fits into the periodic table," explains Schädel. "We are trying to find out if there are significant differences from its lighter homologs—niobium, tantalum, and protactinium—in Group V."

The researchers use a gas jet system to transport the element, one atom at a time, to a tiny cation-exchange-resin column. They have developed an automated rapid-chemistry apparatus for fast, repetitive high performance liquid chromatographic separations of element 105 in aqueous solutions.

"We have shown that the element behaves like a Group V element. For example, we have found that, when dissolved in dilute acid, it is pentavalent," says Schädel. "However, we have also found significant differences in its detailed behavior. In concentrated hydrochloric acid, it forms oxyhalide complexes like niobium and protactinium and not pure halide complexes like tantalum."

The teams are currently preparing an experiment to study the chemistry of element 106. "In general, the element should behave like the Group VI elements tungsten and molybdenum, but theoretical predictions suggest some differences in the detailed chemical properties," says Schädel. ∎

Questions

1. How are transfermium elements synthesized?

2. When element 111 was synthesized during an 18-day experiment in 1994, how many atoms of the element were prepared?

3. For which health problem has heavy-ion therapy been proposed?

Answers begin on page 161.

7 *A very rewarding aspect of chemical research is that sometimes a simple discovery can have profound implications. Detecting the presence of nitric oxide (NO) within the body may be that type of discovery. Ironically, NO has been studied for years as one of the culprits in smog and a contributor to the formation of acid rain. Naturally occurring NO was first noticed in the body when toxicologists started exploring how smog wreaks havoc in the body. Surprised researchers soon discovered that NO is produced by cells all over the body. Recent evidence suggests that this simple compound may play extremely important roles in body chemistry, such as keeping blood pressure in check and boosting the immune system.*

The New Miracle Drug May Be—Smog?

Randi Hutter Epstein

Nitric oxide figures in many diseases, and possible cures.

Eight years ago, at a medical meeting in Rochester, N.Y., two American scientists suggested that human blood vessels spew out nitric oxide gas—the very same toxic chemical in car exhaust fumes and smog. They were greeted by near-unanimous skepticism. But not from Dr. Salvador Moncada, head of British research for Wellcome PLC. To him, the notion suggested a tantalizing path for drug development. He immediately phoned his London team, insisting on crash confirmation.

Back at Wellcome, Moncada's crew started sleuthing for NO in blood vessels with crude environmental-lab machines that measure toxins in car fumes. Put to the test, blood vessels showed enough evidence of NO to persuade Wellcome to make its own souped-up NO detector—1,500 times as sensitive as the others. "Every time we stimulated cells, they released NO," says Moncada. He was convinced this "simple and beautiful" substance lurked all over the body.

In the years since, NO—not to be confused with nitrous oxide (N_2O), or laughing gas—has been found everywhere in the body and tied to an array of major diseases. Today, researchers believe either boosting or blocking NO production could help combat such killers as heart disease and cancer, and treat such ailments as impotence, asthma, and rheumatoid arthritis. Those add up to a colossal potential market, considering the 1.2 million people in the U.S. alone who are diagnosed annually with cancer, the 1.5 million who have heart attacks, and the additional 500,000 who suffer strokes.

It's no surprise, then, that major drugmakers have begun to chase NO-based cures. Thanks to Moncada's prescience, Wellcome is exploring NO treatments for coronary artery disease, asthma, and rheumatoid arthritis, plus a slew of neurological diseases. Ciba-Geigy and Cassella, a subsidiary of Germany's Hoechst, are targeting cardiovascular drugs, while giant Merck is studying NO's role in the brain, the heart, and the immune system. Meanwhile, at least three U.S. startups, Guilford Pharmaceuticals, NitroMed, and Apex Bioscience, have pinned their futures on NO-based treatments for everything from skeptic shock and AIDS-related dementia to brain disorders. The payoff is likely to "be significant, considering the nature of the diseases involved," predicts Dr. Trevor M. Jones, director-general of the Association of the British Pharmaceutical Industry.

Plenty of people think he's wrong. "It's a revolutionary concept" for drug development, agrees Dr. Jeffrey M. Drazen, chief of pulmonary medicine and a NO investigator at Harvard Medical School. "But if I bought stock in drug companies, and I don't, I probably would not be heavily invested in nitric oxide." He worries that while NO can have important beneficial effects, such as lowering blood pressure, it may prove to have toxic effects, too. "We don't know if this is

Reprinted from the December 5, 1994, issue of *Business Week* by special permission. Copyright 1994 by The McGraw-Hill Companies.

another thalidomide," he warns. Besides, since drugmakers need to solve a few problems—targeting the right amount to treat specific conditions, for instance—it is likely to be a decade before significant drugs come to market.

Tickling Veins
Still, it's clear that NO is no longer viewed as merely a pollutant. Toxicologists began to change the view of it in the 1980s, when they started exploring how smog wreaks havoc in the body. They spotted excess NO, way beyond the amount that would be expected from pollution—the first clue that the body might be making its own supply. Then, immunologists led by University of Utah School of Medicine's Dr. John B. Hibbs, Jr. zeroed in on NO in macrophages, an immune-system cell, which suggested that NO can fight infections and possibly tumors. About the same time, Louis J. Ignarro, professor of molecular pharmacology at the University of California, Los Angeles School of Medicine, and Robert F. Furchgott, professor emeritus of pharmacology at the State University of New York Health Science Center at Brooklyn College of Medicine, proposed that wisps of NO tickle the inner lining of blood vessels, prompting them to relax. That was the finding that sent Moncada racing to the phone. Any substance that could do that, he knew, could be valuable because so many diseases are tied to narrowed blood vessels, including cardiovascular disease.

NO's ubiquity is both an asset and a drawback. "With the involvement of this molecule in so many functions, the risk is that [drugmakers] may be working in the wrong area," says Duncan Moore, a health-care analyst at Morgan Stanley & Co. in London. Indeed, to develop successful NO-based treatments, scientists first have to figure out NO's role in a disease, whether it's helpful or harmful, then figure out how to take advantage of or avoid the effect. Then they need to deliver NO precisely where it's needed, and in just the right dose.

Good News, Bad News
The effort to treat impotence shows how difficult this can be. Many scientists think the condition is caused by lack of NO in the nervous system. In initial attempts in 1992, University of California at San Francisco researchers injected NO into the penis to relax blood vessels and get blood rushing to the site. The men did get erections, but they also fainted because this first attempt used long-lasting NO drugs that lowered blood pressure.

Despite such obstacles, drugmakers are forging ahead. Wellcome, NitroMed, and Apex are targeting septic shock, potentially fatal low blood pressure caused by massive infections that generate excess NO. Roughly 600,000 people in the U.S. develop sepsis each year, and 100,000 die. In recent years, three high-profile potential treatments have failed. Analysts say a drug that helps could bring in as much as $1 billion a year in the U.S. alone.

In September, Wellcome launched clinical trials for its drug, L-NMMA, which inhibits enzymes that make NO. Meanwhile, Apex Bioscience Inc. in Durham, N.C., has won a patent for a potential septic-shock therapy. Its goal is to use a recombinant hemoglobin that would mop up excess NO caused by infections. In Cambridge, Mass., NitroMed researchers are gambling on a different approach. Their patented method aims to block NO receptors on cells to keep the chemical at bay. None of these drugs would cure the underlying infection, but they might restore blood pressure, keeping patients alive until the original illness can be controlled, says Joseph DeAngelo, vice-president of research at Apex.

Moncada's 1986 phone call was crucial in unleashing NO research. Today, he insists, research into the workings of NO "is moving faster than any other field of research." Still, it will be a while before NO becomes more than just smog. ■

Questions

1. In what way does nitric oxide assist the circulatory system?

2. What impact does NO have on the immune system?

3. What effect does the drug L-NMMA have on NO production?

Answers begin on page 161.

8 *By compressing ordinary compounds under millions of atmospheres of pressure, scientists are breaking down and reforming chemical bonds, creating new and exotic materials. The exploration of extremes in pressure has resulted in oxygen crystals with red, yellow, and blue facets, superhard glass from quartz, and potential superconductors from compressed hydrogen. Under intense pressure, atoms are compressed together so tightly that electrons change orbitals and form new bonding patterns. Most materials then collapse into dense new substances. With an eye toward commercialization of high-pressure materials, researchers will create many new substances with undreamed of properties. Their studies may also provide clues to the materials and compression processes deep inside the earth.*

The Big Squeeze in the Lab: How Extreme Pressure Is Creating Exotic Materials

Ruth Coxeter, with Peter Coy

In 1987, C.W. (Paul) Chu of the University of Houston was racing scientists from Moscow to Long Island for a breakthrough in high-temperature superconductivity. On a hunch, he prepared a bit of copper, barium, lanthanum, and oxygen smaller than a BB. Then he worked high-pressure alchemy, squeezing it under pressure 18,000 times as great as that at sea level.

Bingo: When electrodes were plugged in to the blend, it conducted electricity with zero resistance. Chu then guessed he could achieve the some effect at ordinary pressure by substituting yttrium for lanthanum. The substitution worked—and earned Chu worldwide fame for a material that superconducted at a record high temperature: at −180C.

Chu's work highlighted how extraordinarily high pressures can strong-arm elements into yielding their secrets. The high-pressure manufacture of synthetic diamond (by squeezing carbon) is already a near-billion-dollar industry. Today, researchers are using pressures exceeding those at the center of the earth to break and re-form chemical bonds in everything from hydrogen to beach sand. "Even by a conservative estimate, we could triple the range of materials we now know just by squeezing them," says Robert M. Hazen, a research scientist with the Carnegie Institution's Geophysical Laboratory and author of a 1993 book on the topic, *The New Alchemists*.

If gases can be compressed into stable metallic crystals they might superconduct even at room temperatures. Superhard glass made from silicon dioxide—or quartz—could serve as windshields for rockets. A metallic form of hydrogen could be an incredibly dense form of stored chemical energy. "It would be something like 30 times more efficient than any existing rocket fuel," Hazen says.

The laboratory equipment required to achieve these awesome pressures, a diamond anvil, is no more than 8 inches tall and can be screwed tight by hand. It works by amplifying arm power through gears and concentrating all of it into an extremely small area. That's done by placing a tiny sample between the tips of two cut diamonds. A plate with a hole in it corrals the sample.

Research on such a small scale has big drawbacks. The products are too tiny—about a millionth of an ounce—to be useful for anything except research. Many materials revert to normal once the pressure is off. And high-pressure science can be hazardous. Having witnessed one explosion, David Mao and Russell Hemley of the Carnegie Institution now operate experiments from behind a steel wall.

With real money probably years off, most work on exotic high-

Reprinted from the February 13, 1995, issue of *Business Week* by special permission.
Copyright 1995 by The McGraw-Hill Companies.

pressure materials is going on in universities, not companies. Cornell, Harvard, and the University of California at Berkeley have substantial programs. Three years ago, the National Science Foundation established a Center for High-Pressure Research, pooling efforts at the Carnegie Institution, Princeton, and the State University of New York at Stony Brook.

Hard Luck
High-pressure research got a jolt last May when 11 Russian and French researchers claimed in the journal *Physics Letters* that they had created a material harder than diamond from "buckyballs," or carbon-60. But similar claims have arisen before, and some scientists argue that the samples were too small to be reliably tested.

Squeezing materials does more than harden them. Compressing gases such as hydrogen liberates their electrons, turning them into excellent electrical conductors. Helium, ordinarily inert, will bond with nitrogen when sufficiently compressed. Oxygen is ordinarily colorless—but under extreme pressure it forms crystals with facets of red, yellow, and blue. UCLA professor of physical chemistry Malcolm F. Nicol speculates that sulfur crystals could store information, with different colors of light serving to write, erase, and read data.

The quest for a superconducting metallic hydrogen is taking scientists into realms of pressure never experienced on earth—and into heated rivalries as well. Researchers at Harvard University and Carnegie Institution disagree over what happens to hydrogen at 1.5 million atmospheres, or megabars, at which point its molecules line up like watermelons. Carnegie brags it was the first to reach 2.5 megabars. No one has surpassed 3 megabars, where scientists speculate hydrogen becomes fully metallic and superconducting. The big question: Will metallic hydrogen remain metallic once the pressure is off?

Micropops
Commercialization is just one goal. High-pressure researchers also seek to understand the composition of stars and planets and the sources of earthquakes. UCLA's Nicol has even recreated conditions of the early earth to form chains of hydrogen cyanide, which he believes are precursors to DNA. One favorite for research is ikaite, a watery calcium compound. Its behavior under high temperature and pressure sheds light on how veins of gold, silver, and other ores are formed. Raymond Jeanloz, a physicist at UC Berkeley, and graduate student Charles Meade proposed a new seismic theory after pressurizing serpentine, a greenish, waxy mineral. A microphone attached to a diamond anvil picked up a popping sound—perhaps a miniature version of a quake that occurs when water is forced out of serpentine deep underground.

Doubts about the commercial value of exotic high-pressure materials have kept away mainstream producers of synthetic diamonds, such as General Electric, Sumitomo, and De Beers. "We're a materials supplier. Just creating high-pressure phases doesn't make you any money," says William F. Banholzer, manager of engineering for GE Superabrasives in Worthington, Ohio. But in labs like Paul Chu's, the big squeeze is paying off. Chu recently found a superconductor that he hopes could be produced in industrial quantities by depositing the materials in vapor form. That's the kind of advance that keeps the pressure on. ■

Questions

1. In generating extremely high pressures, a special anvil is used. What material is used on the tip of the anvil?

2. How might a metallic form of hydrogen be useful?

3. How does oxygen change when placed under extreme pressure?

Answers begin on page 161.

9 *Chemistry often plays a central role in the preservation and restoration of historic artifacts. Objects recovered from the* Titanic *show significant deterioration even though at the wreck site there is no light and the temperature is near 0°C. Wooden items are devoured by shipworm mollusks. Marine bacteria feed on metals, producing black sulfides that degrade and stain artifacts. Metallic objects also undergo corrosion by electrochemical processes. After metal objects are recovered and analyzed to determine the identity of alloys, they are cleaned and stabilized using electrolysis, a chemical process driven by the application of electrical energy. This technique involves passage of a weak electrical current through the artifact, which loosens surface corrosion and draws out chloride ions that would cause future corrosion.*

Chemical Techniques Help Conserve Artifacts Raised from *Titanic* Wreck

Michael Freemantle

Compexing agents and electrolysis restore some artifacts from the* Titanic, *providing insights for research in deep-sea areas.

Chemistry has played a central role in the restoration of more than 150 artifacts on display at an exhibition titled "The Wreck of the *Titanic*," which opened this month in London. Scientific examination of metal debris, corrosion products, and artifacts removed from the wreck site may help in the management of other maritime archaeological sites and provide key information for long-term storage and containment under seawater.

The exhibition at the National Maritime Museum in Greenwich features the first major display of artifacts recovered from the seabed around the wreck of the "unsinkable" ocean liner in the icy North Atlantic Ocean. Exhibited relics include a porthole, a chandelier, buttons from crew uniforms, a pipe, and a leather cigarette case complete with cigarettes.

Among the more delicate relics retrieved from the wreck are paper items found tightly wadded in leather bags. They include a newspaper published in Southampton on Tuesday, April 9, 1912, the day before the *Titanic* sailed from port on her maiden voyage. The state of preservation of paper recovered from the seabed is unparalleled in any other shipwreck excavation.

Stephen Deuchar, head of exhibition and display at the museum, points out that "state-of-the-art technology, as complex and advanced as space technology," has enabled the wreck to be discovered and explored some 2½ miles under the ocean.

Three major recovery expeditions have made their way to the *Titanic* in the submersible *Nautile*, a craft designed for deep-sea scientific research with a crew of three. The submersible, a titanium-hulled bathyscaphe, withstands a pressure of about 400 atm at the wreck site. *Nautile* can move about under its own power on the seabed while maintaining contact with the survey ship *Nadir* at the surface. The submersible's robotic arms and a video camera mounted on a remote-operated vehicle have enabled a wide range of material to be collected from the site.

On the ocean bed at the wreck site there is a complete absence of light, a temperature of around 0 degrees Celsius, and very little oxygen—conditions that have slowed the deterioration of objects. Nevertheless, many of the artifacts have suffered significant biological and chemical degradation as well as erosion by sand and currents.

For example, shipworm mollusks devour wooden items as they burrow through them. Bacteria on the seabed feed on metals, especially iron, producing black sulfides that degrade and stain many artifacts. The fibers of natural organic materials break down and separate making them porous and fragile. Metallic objects undergo corrosion, the effect of an electrochemical process.

In 1991, Toronto-based IMAX Corp. charted the Russian Academy of Science vessel *Akademik Keldysh* to film the wreck. During one dive, the manned submersible *MIR-2* recovered nine metal fragments of the *Titanic* and samples of corrosion products and sediment from the site.

Patrician Stoffyn-Egli, a researcher at Microchem, Geochemistry Consultants, East Jeddore, Nova Scotia, and Dale E. Buckley, a scientist with the Geological Survey of Canada, Bedford Institute of Oceanography, Dartmouth, Nova Scotia, examined the samples recovered by the Russian ship with a power x-ray diffractometer and a scanning electron microscope (SEM) equipped with an energy dispersive x-ray spectrometer. They aimed "to gain an insight into the geochemistry of iron and other metals that had accidentally been introduced into the deep-sea environment."

Their study shows that bacteria play a major role in promoting corrosion of the *Titanic*, resulting in fast-growing structures such as rust flows and rusticles. Rusticles resemble stalactites and grow on the hull of the ship. They have a brittle hydrous iron oxide shell with a dark-red outer surface and an orange inner surface. The core of the rusticle and the inner surface of the shell consist of spherical aggregates of small needlelike crystals of goethite (α-FeO(OH)). The outer surface of the shell has the crystal structure of lepidocrocite [γ-FeO(OH)]. The Canadian study shows that a variety of bacteria are present in the rusticles. These are predominantly sulfate-reducing species that multiply rapidly in anaerobic conditions.

The rust flakes contain a mixture of goethite, lepidocrocite, and other minerals. SEM examination of the flakes reveals a well-crystallized iron mineral, possibly hematite (Fe_2O_3), associated with a silicon-rich iron mineral. Black patches of siderite ($FeCO_3$), and iron-rich cubes, possible magnetite, also are present. Surprisingly, the rust flakes have a thin coating of lead carbonate ($PbCO_3$). Small cubes of galena (PbS) also are common. The Canadian researchers suggest that the source of the lead is paint on the hull of the *Titanic*.

Buckley tells C&EN that his research team has been "continuing some of the research on the *Titanic* material using a new environmental scanning electron microscope coupled with a new energy dispersive x-ray spectrometer."

Objects from the wreck absorb chlorides and sulfates from the seawater and become weakened, stained, and encrusted. It is therefore essential that when they are raised to the surface they are kept wet. Ceramic and glass items can lose surface layers if salts are allowed to crystallize. Leather can harden, crack, and shrink as it dries out if salts in the pores are not removed before drying. Metallic objects undergo accelerated corrosion if exposed to oxygen.

The levels of corrosion of metal objects raised from the site vary immensely as a result of "the microstructure of the metal and its in situ environment," according to conservator Stéphane Pennec, head of the LP3 Conservation Laboratory in France that is currently restoring objects recovered during the 1993 and 1994 expeditions by the *Nautile*. "Some metal objects, such as bronze whistles, and nickel pots, look brand new," observes Pennec. "Other objects made of copper alloys are nightmares because they are so mineralized and fragile."

As soon as artifacts are brought up to the surface, acidic silt is washed off. The artifacts are then examined and numbered before being packed into tanks of water. "One of the major problems is packing the objects in the water," Pennec says. "It may be several weeks before they arrive at the conservation laboratory ashore. If the boat moves a lot they can easily be damaged. We separate the artifacts according to material types and pack them with foam to minimize movement."

On arriving at the laboratory, the artifacts are washed with fresh water and, if the objects are not too fragile, cleaned with a brush to remove corrosion products. With organic materials, "the main trouble is fungi and bacterial growth," Pennec says. "So, if we need to store them for a while, we use a biocide in the water."

Surface salts are washed out with frequent water changes, and the salts trapped in an object are removed by electrolysis or treatment with complexing-reducing agents. "To reduce corrosion of iron objects we use an alkaline dithionite bath, and for desalination of copper objects we use sesquicarbonate [a mixture of hydrated Na_2CO_3 and $NaHCO_3$]. Our main aim is the removal of chloride because it promotes corrosion of metals and depolymerizes organics," says Pennec.

Seawater is slightly alkaline, with a typical pH of about 8.2. However, the pH of the microenvironment around an iron artifact immersed in seawater can fall to as low as 4.2. This arises from the hydrolysis of metal ions forms by the oxidation of iron:

$$Fe \rightarrow Fe^{2+} + 2e^-$$
$$Fe^{2+} + 2H_2O \rightarrow Fe(OH)_2 + 2H^+$$

As corrosion proceeds, chloride ions from the surrounding seawater diffuse to the corroded metal surface.

Chemist and conservator Ian D. MacLeod, who has worked closely with Pennec on several conservation projects, observes that accelerated corrosion of iron in a chloride-rich acidic environment results in the loss of much of the archaeological value of an artifact. MacLeod, who is head of the department of materials conservation at Western Australian Maritime Museum, Fremantle, points out that copper-based materials such as bronze and brass corrode to form copper hydroxychlorides. Silver objects corrode to form silver chloride, although the rate of corrosion is much slower than for copper-based materials.

"One thing that surprised me with the *Titanic* artifacts was the way in which silver plate was totally disbonded from the copper base," says MacLeod. "Unless you are extremely careful, the original silver plate, which is quite a thick layer, will just float off the surface. Normally, you get the metal underneath pitting and pustules growing up through the plate—but here, perhaps because of the pressure, you get the corrosion products underneath lifting off the solid silver plate which remains remarkably intact."

SEM and x-ray analyses of many of the objects are carried out to determine the composition of alloys and corrosion products. On the basis of this information, it is possible to determine the best treatment for the relics.

Electrolysis is used to clean and stabilize strong metal artifacts, including most of the bronze objects. The electrolysis cell consists of the object itself, which is the cathode, a stainless steel anode, and an alkaline electrolyte, typically a dilute aqueous solution of sodium hydroxide or sodium carbonate. The passage of a weak electric current draws out the chloride ions from the metal. When a stronger potential is applied, the hydrogen bubbles produced at the cathode bubble over the surface of the object and loosen corrosion products and any concretion. Concretion is a mixture of calcareous deposits.

"Electrolysis does two things," says MacLeod. "It stops further corrosion, and it also removes the chloride from deep within the heart of the object. That means, after treatment, when the analysis shows there are no more salts coming out, you take it out, dry it, and put a protective coating such as a clear wax on it."

Some ceramic artifacts and organic materials are cleaned by electrophoresis. The artifact is placed between positive and negative electrodes and a potential applied. Salts, dirt, and other particles located within the electric field break up and their charged components migrate through the solution to the electrodes. The method is used to remove contamination from bank notes, wallets, and leather bags. However, some materials are too delicate for this type of treatment.

An electrolytic pen also is used to remove stains from thick ceramic objects. The pen is a miniature electrolytic cell. The object to be cleaned is either dipped into an electrolyte or the part with the stain is placed under an electrolyte drip. The pen moves around on the object to clean areas much larger than its own diameter.

In Western Australia, MacLeod has developed a new method for treating organic materials that "works brilliantly with newspaper." The method is employed by Pennec to conserve some of the *Titanic* artifacts. "The new method uses a complexing agent in neutral solution, for example a neutral citrate solution containing sodium dithionite, which is an additive regularly used in the oil industry to scavenge oxygen," explains MacLeod. "This reduces the oxidation state of the rust lattice or the iron corrosion product from iron(III) to iron(II). The lattice then becomes unstable, and so it begins to fall apart. Because you have a neutral buffered citrate ion present, you get a hydrous iron oxide citrate complex formed that is very stable."

The method removes iron stains and contamination from delicate textiles or newspapers, "so you do not use aggressive solutions like strong acids to dissolve the inert corrosion products," MacLeod adds.

Pennec also uses oxalic acid as a complexing agent to remove corrosion products. The acid complexes with iron oxide to form iron oxalic, which can be readily removed by rinsing with water.

Of all the materials, glass is most resistant to deterioration and many of the glass items recovered from the wreck site are complete. One of the relics on display at the Greenwich exhibition is a champagne bottle with its original contents. For glass objects, surface deposits are removed with scalpels or chemically. The main problem with glass is surface iridescence caused by hydrolysis of the silica network.

Disfiguring concretions on ceramic items are removed by hand and soluble salts are washed out by rinsing with water. Deep-penetrating iron corrosion stains are removed by electrolysis or a neutral solution of complexing agent. Leather and wood need a consolidant to replace the material consumed by microorganisms. This helps to minimize cell collapse.

Following conservation, the artifacts are carefully preserved. "The storage conditions have to be just right," says Gillian Hutchinson, curator of archaeology at the maritime museum. She explains: "You have to have the right temperature and humidity, particularly with bronze- and silver-plated artifacts where corrosion can start flaring up all over again if they go into a damp environment. Paper items have to be kept out of bright light, otherwise the ink fades. Provided this is done, artifacts will survive almost indefinitely, whereas artifacts still down there on the seabed will continue to deteriorate."

The information gained from scientific examination of the wreck of the *Titanic*, the debris field, and the artifacts, is providing valuable insights for marine engineers and archaeologists. "By studying a range of shipwrecks in a range of sites and conditions, we can understand more about long-term deterioration rates of materials," says MacLeod. "This type of data is essential if we are going to monitor and maintain undersea gas pipelines, long-term containment of radioactive waste, and nuclear submarines that have sunk."

MacLeod cites as an example the wreck of the *Batavia*, which sank off the coast of Australia in 1629. His work on the wreck revealed a bronze alloy of unusual composition that is exceptionally resistant to corrosion. A similar composition was chosen by a water authority in New South Wales for a main sewage outfall pipe," notes MacLeod.

"It's absolutely impossible in any laboratory situation to mimic the dynamics of the marine environment—that is, the interaction of your object with the sediment, the seawater, and the microflora," concludes MacLeod. "We really need to go down onto the site of the *Titanic* and conduct in situ measurements to obtain the end results of this beautiful long-term corrosion experiment."

Pennec and MacLeod are hoping to participate in another expedition to the wreck of the *Titanic* next year. ■

Questions

1. Why must some items be kept wet after they are recovered from the ocean?

2. List two ways in which recovered artifacts benefit from electrolysis.

3. Which material is most resistant to deterioration and corrosion?

Answers begin on page 161.

10 *There is widespread concern that the greenhouse gases we are pumping into the atmosphere—carbon dioxide, methane, and chlorofluorocarbons are the main ones—are warming the globe and changing the climate. The combustion of fossil fuels, which produces carbon dioxide (CO_2), accounts for a vast amount of all of the energy consumed in the United States. It is this enormous emission of CO_2 that threatens to change the world's climate through the greenhouse effect. In a greenhouse, the sunlight passes through the glass panels and is transformed into heat, which is trapped inside the building. The earth's atmosphere behaves in a similar way, preventing heat from radiating into space. A promising approach to easing the problem of CO_2 buildup is the development of chemical processes that consume CO_2 as a feedstock.*

Carbon Dioxide: Global Problem and Global Resource

Kimberly A. Magrini and David Boron

Global emissions of carbon dioxide are of increasing environmental concern. The problem could be eased by using the gas as a resource in chemical processes.

Given that atmospheric carbon dioxide, methane and other greenhouse gas levels are rising, concern about their climatic effects continues to grow. Such climate changes could seriously affect natural ecosystems, agricultural production and ocean levels. Controlling these emissions, however, will directly involve energy, economic and development policies worldwide. Since these uncontrolled emissions pose potentially serious threats to the environment, several feasible strategies have been identified by global consensus that focus on CO_2 mitigation and use.

Routine atmospheric monitoring, begun in 1958 and significantly expanded during the 1980s, has produced a reasonably comprehensive picture of trends in atmospheric CO_2 levels in the recent past. Analysing air trapped in polar ice cores has extended such information back about 100,000 years. Based on these analyses, it is apparent that average global atmospheric CO_2 concentrations have increased from pre-industrial levels of around 270 parts per million by volume (ppmv) to a 1991 value of 355 ppmv. A comprehensive monitoring programme for CO_2 conducted by the US National Oceanic and Atmospheric Administration (NOAA) shows that the average global growth rate for atmospheric CO_2 from 1981-1991 is 1.55 ppmv/a.[1]

The modern-day build-up of CO_2 in the atmosphere is believed to be caused by an excess of man-made emissions over and above that taken up by natural CO_2 sinks such as the oceans, forests and crops. Major man-made sources of CO_2 include fossil fuel combustion and deforestation. Globally, industrial CO_2 emissions have tripled since 1950 and are currently about 22×10^{12} t/a in the atmosphere.

All regions of the world show steadily increasing CO_2 emissions from 1950, with more developed North American, European and East European regions showing the highest levels. Although more recent trends indicate that the North American and European emissions appear to be levelling off, while emissions from the developing areas are steadily increasing, the more developed regions still account for 70% of the global total.

As with CO_2, recent measurements of atmospheric methane levels along with polar ice core analysis show that they have also risen significantly since the onset of the industrial era. Current concentrations average 1.7 ppmv, more than double the pre-industrial value. Encouragingly, NOAA data show that, although atmospheric methane levels are still increasing, the rate of increase has declined from

13.5 parts per billion by volume (ppbv) in 1983 to about 9.5 ppbv in 1991.

Natural methane emission sources include wetlands and the oceans. Industrial sources are represented by coal mining, sewage treatment, natural gas production, rice cultivation, landfills, and fossil fuel and biomass burning. According to a Stockholm Environment Institute report,[1] rice cultivation and livestock account for 60% of man-made methane emissions to the atmosphere. Energy-related sources contribute 23% with China, the US and the former Soviet Union accounting for 60% of this amount.

The United Nations (UN) Convention on Climate Change, an agenda of the UN Conference on Environment and Development (Rio Conference), developed a framework that begins to address the stabilisation of these emissions at levels that will prevent future human activities from dangerously interfering with the global climate system.[2] Some of the measures include preparing databases on greenhouse gas emissions and sinks; adopting policies that address climate change; developing, using and transferring technologies that directly reduce emissions; and sustainably managing natural sinks.

The measure addressing technology development to reduce emissions directly applies to using atmospheric CO_2. Since significantly more CO_2 than methane is present in the atmosphere, CO_2 provides a ubiquitous if dilute source of carbon, which can be used for producing fuels, chemicals and materials. Another benefit of using CO_2 lies in pollution reduction or prevention by substituting CO_2 as a carbon source for more toxic chemical feedstocks currently being used by the chemical industry.

It is interesting to compare the estimated total amount of atmospheric carbon present as CO_2 (10^{12}t) with the amount of carbon needed if all chemical and fuels produced in the US in 1993 were made from CO_2 (about 10^{10}t).[3] Another estimate places global fossil fuel emissions in 1989 and 1990 at 6×10^9t and atmospheric CO_2 at 7.5×10^{11}t.[4] Taken together, these values show that CO_2 use is unlikely to function as a primary CO_2 sink. Usage will, however, begin to provide a loop to recycle CO_2, which will in turn reduce a portion of atmospheric CO_2 emissions. Successful research and development efforts will provide new technologies, stimulate new business, and, in turn, increase economic growth.

The impact of successful CO_2 use would be felt most directly by the chemical industry which would use new chemical and biochemical processes for producing high-value chemicals, and at the same time avoid potential environmental penalties for CO_2 emissions. Direct technology benefits include a better understanding of CO_2 chemistry and biochemistry that may lead to new products and materials as well as fuels and chemicals. Economic benefits comprise increased technological competitiveness for industry, and the ability to use CO_2 as a feedstock for chemical production instead of current petroleum-based feedstocks, which may become scarce or environmentally less acceptable in the 21st century.[5]

A conservative estimate indicates that CO_2 usage could easily be expanded by a factor of ten or more if new or improved processed could be developed. Currently, chemical companies throughout the world are developing new approaches for using CO_2. In an industrial collaboration in Japan, Mitsubishi Heavy Industries is looking at producing algae-based biomass from CO_2-containing power plant stack gas.[6] Another example is a laboratory scale Japanese process which directly converts CO_2 to gasoline in a two-stage flow through catalytic reaction.[7] Several countries have recently proposed carbon taxes that would make CO_2 recovery and use attractive whenever it is feasible.

Although CO_2-based processes are currently in use, problems with further development do exist. By itself CO_2 is fairly inert. Activating it so that it will react requires significant energy input. Producing fuels and chemicals from CO_2 will require an inexpensive and environmentally acceptable energy source. However, higher-value products such as speciality chemicals may be able to absorb higher feedstock costs while remaining competitive. For example, hydrogen generated electrolytically, which appears expensive to use in converting CO_2 into fuels, may be reasonably priced feedstock for producing a high-value chemical such as a plastic.

Since global warming is the key issue with respect to CO_2, why then develop processes that can only use a small fraction of the available CO_2? One motive is that profits generated from successful use can be applied to offset disposal costs, while concurrently developing the technology that may lead to revolutionary new products. As profit is the primary goal, the scale of CO_2 usage is not a major issue. However, technical expertise is: expertise is required both to prod CO_2 into reacting and successfully position the resultant CO_2-based products for replacing entrenched products.[8]

Other compelling reasons for pursuing CO_2 usage research include the availability and purity of the resource; the experience the chemical industry already possesses in using CO_2 as a feedstock; the advancement of international research efforts which may lead to industrial dependence on foreign technology; and environmental concerns about continued emissions, toxic feedstock use, and the levelling of potential carbon taxes.

The chemical industry currently uses CO_2 as a feedstock in several commercial processes. Methanol (10Mt/a) is produced from synthesis gas (carbon monoxide and hydrogen) and/or CO_2 using a copper-based catalyst at moderate temperatures and high pressures (250°C and 80atm).[8] Urea, (80Mt/a) used in making resins, polyurethanes, and thermosetting materials, is commonly produced by reacting ammonia and CO_2 over a zinc catalyst at moderate temperature and high pressure. Salicylic acid (25,000t/a) is produced when CO_2 is added to a phenol derivative. Other processes incorporate CO_2 into polycarbonates, a versatile material that is used in compact disc and container manufacturing. All of these processes can be generalised to form similar types of chemicals and materials.

A newly-developed pigment process, which directly uses CO_2, also reduces toxic by-products.[9] This process reacts CO_2 with lime to form precipitated calcium carbonate (PCC) particles, which are used to replace titanium dioxide (TiO_2) as a paper whitener and as a paint pigment. Paper manufacturers and paint producers use substantial amounts of TiO_2, which has a worldwide market of about 3.2Mt/a valued at $6bn.[9] Although TiO_2 is widely used, its high price ($2000/t) and the environmental pollution associated with its manufacture make it an attractive target for replacement with alternative pigments. One advantage of the PCC process is that it avoids forming the toxic by-products associated with TiO_2 synthesis. In addition, $CaCO_3$, unlike TiO_2, can be directly precipitated in the paper fibres, which exhibit stronger wear and abrasion properties.

The kind of CO_2 used as feedstock for the PCC paper whiteners is frequently obtained from flue gas generated by the paper mills which then buy back the PCC. Controlling the size and shape of the calcium carbonate crystals provides a $200-500/t product to compete with the $2000/t TiO_2. The new calcium carbonates have already won a 680,000t/a market position in the US, and growth continues. These and other processes currently use CO_2 at a level of about 100Mt/, although exact figures on the total carbon balance are generally not available.

Biologically, CO_2 has been used to produce fats from algae for biodiesel fuels and the antioxidant vitamin B-carotene. Though algae-derived biodiesel has only been produced in laboratory scale quantities and B-carotene production worldwide is only 10t/a, such biological systems, if developed, have the potential to produce many different oxygenated chemicals from CO_2. Another example of this is polymer synthesis from bacteria. Polyhydroxybutyrate/valerate (PHBV), a polyester, is manufactured by a commercial microbial process using glucose as the carbon source. The microbes use glucose to form polyester pellets of greater than 96% purity. The micron-sized pellets are removed from the bacteria for further processing. ICI produces this plastic and sells it for $8-12/lb. Replacing glucose, an expensive feedstock which only contributes carbon to the microbes, with CO_2 would lower costs and subsequently increase the use of this biodegradable plastic in commercial applications. Recent results demonstrate that these plastics can be made readily from CO_2 alone and various strains of bacteria.[10,11]

Other chemical processes being developed in laboratories include the direct reaction of CO_2 to form polycarbonates with superior properties compared with material produced with current methods.[12] This process uses supercritical CO_2 as both solvent and carbon source and a homogeneous catalyst to form the polymer under relatively mild conditions (80-100°C, 2000-5000psi). The resulting material consists of about 50 mole% CO_2 and a molecular weight of 200,000, which is significantly better than that achievable with currently used chain polymerization methods. The next step in this work will be to use CO_2 to make aliphatic polyesters, an important class of compounds currently made with energy intensive condensation methods that limit the polymer molecular weight that can be obtained. The CO_2-based process provides a low energy route through chain polymerization that is not limited in molecular weight. The challenge in this work lies in developing more active catalysts.

Generally, one major constraint to finding new methods of using CO_2 is catalyst development. Identifying new catalysts for CO_2 conversion has not received much emphasis in comparison with the large quantity of work performed with CO (syngas) catalysts; thus, looking at modifying CO activation catalysts may provide a promising area to begin the search for new CO_2 catalysts. Additionally, some postulated new used of CO_2 include reacting it with α-amino nitriles and water in a two-step process to form hydantoins, a class of materials which exhibit interesting pharmaceutical activity.[12] Another possibility is examining the degradation characteristics and solvent properties of cyclic carbonates which are easily and directly made from CO_2.[12]

A long-standing goal related to CO_2 use is the dimerisation of methane to form the C_2 hydrocarbons ethane and ethylene with efficient yields and selectivities. These compounds are common feedstocks for the production of a variety of chemicals. The minimum ethylene yield for an economically viable one-step methane conversion process has been identified as 40%.[14] Many catalysts that exhibit excellent C_2 selectivity (>90%) have been found; however, yields obtained with these catalysts have not exceeded 2%, and selecting conditions that favor higher methane conversion uniformly produced dramatic decreases in C_2 selectivity. This low selectivity at high methane conversions is attributed to the higher reactivity of the C_2 products with oxygen when these product concentrations approach that of the methane feed. Dramatic improvements in C_2 yield were recently obtained when these products were partially separated and removed from further reaction by trapping them on adsorbents such as molecular sieves. The products were subsequently recovered by thermally desorbing the trap. An ethylene selectivity of 88% was achieved with 97% methane conversion.[13,14]

This type of reaction system, in which highly reactive desired products are protected through selective adsorption, should be applicable to other reactions such as the partial oxidation of methane to formaldehyde or methanol with the appropriate choice of catalyst and adsorbent. Such systems may also apply to CO_2 conversion processes.

Currently, the efficient direct synthesis of gasolines from CO_2 has not yet been achieved industrially because (1) conventional Fischer-Tropsch catalysts selectively form methane at the conversion levels required for efficient gasoline production; (2) conventional methanol catalysts do not work as well for CO_2 hydrogenation as they do for syngas conversion; and (3) the combination of methanol synthesis catalysts with H-ZSM-5 zeolite catalysts produces only gaseous paraffins rather than gasoline hydrocarbons because of the high hydrogenation activity of the zeolite.

Some of these challenges have been addressed in preliminary work by Tomoyuki Inui of Kyoto University,[7,15] who uses a two-stage catalytic process to convert CO_2 to gasoline successfully with a one-pass, series-connected reactor. In the first-stage reactor, CO_2 was converted to methanol over a Cu-Zn-Cr-Al/mixed oxide catalyst. The total reaction mixture exiting the first stage was fed into the second-stage reactor containing a protonated H-Fe-silicate pentasil-type catalyst which produced gasoline with a 50% selectivity. The other products were light olefins which can be circulated back through stage two for further conversion to gasoline.

Monsanto is developing a process in which CO_2, instead of phosgene, reacts with amines to form urethanes, an important class of materials. This process, undergoing pilot-scale testing, reacts CO_2 with amine compounds to form carbamate ions, which are then reacted with allylic chlorides in a mild, palladium-catalyzed reaction that selectively forms allylic urethanes with high yields (66-100%).[16] Perhaps the primary benefit of this new process is replacing phosgene with CO_2 as a starting material for urethanes, typically made by reacting an alcohol with an isocyanate. The isocyanate is obtained by adding phosgene to an amine with HCl forming as the major by-product. Thus, substituting CO_2 for $COCl_2$ replaces a toxic feedstock and eliminates a hazardous waste stream. Another benefit is that the CO_2-based process allows the introduction of an allylic group into the urethane, giving a product containing a reactive group with which additional chemistry can occur. Allylic urethanes have been successfully used in radiation-curable coating systems, as organic synthesis intermediates and in materials for shatter-proof applications such as glasses, face-shields and windows.

An economic analysis of using CO_2 to replace phosgene shows that the new process must be 30-40% more efficient to be competitive because phosgene is a cheap feedstock. However, the environmental benefits of such replacement, though difficult to price now, may become considerable in the near future, especially if carbon taxes and chlorine bans are enacted.

Chemical processes based on CO_2 have advantages and disadvantages. The advantages include CO_2 being cheap and abundant; potential energy saving through CO_2 recycling; avoiding or reducing toxic feedstock use; less severe reaction conditions; and the possibility of discovering completely new and useful chemistry. The disadvantages are the current low conversions rates, yields and selectivities achievable, high energy requirements and the need for better catalysts and economical hydrogen sources. Other challenges must include the ability of the chemical industry to use coal rather than CO_2 as a feedstock when petroleum supplies become scarce, and the ease of re-engineering current processes that address environmental or feedstock concerns rather than adopting new ones based on CO_2.

In summary, CO_2 emissions to the atmosphere continue to be a growing concern with respect to global climate change. The amount of CO_2 released yearly is such that a chemical and fuels industry based entirely on CO_2 will use less than 1% of the total emissions. However, the economic and environmental benefits of a CO_2-based industry could be considerable. Avoiding the use of toxic feedstocks, future CO_2 emissions and hazardous by-products are compelling environmental drivers. Already, CO_2 has been incorporated into pharmaceuticals, fuels, chemicals, and into known and new materials of new or better qualities.

References

1. 'Environmental data report 1993-1994', United Nations Environmental Program, GEMS Monitoring and Assessment Research Centre, World Resources Institute, UK Dept of the Environment, London, 1993, 8-10
2. *Ibid*, 380-387
3. Harrymore, B., personal communication, 1994
4. Houghton, R.A., *Environ. Sci.*

Technol., 1990, **24**, 414-22
5. Volintine, B., *et al*, Internal DOE report, April 1993
6. Nishikawa, N., *et al*, *Energy Convers. Mgmt.*, 1992, **33**, 553-60
7. Inui, T., Takeguchi, T., Kohama, A., & Tanide, K., *ibid*, 1992, **33**, 513-20
8. Aresta, M., Quaranta, E., & Tomosi, L., *ibid*, 1992, **33**, 495-504
9. Lipinsky, E.S., *ibid*, 1992, **33**, 505-12
10. Kerns, R., personal communication, 1993
11. Maness, P.C., & Weaver, P.S., *Appl. Biochem. Biotech.*, 1994, **45/46**, 395-406
12. Beckman, E., personal communication, 1994
13. Tonkovich, A.L., Carr, R.W., & Aris, R., *Science*, 1993, **262**, 221
14. Jiang, Y., Yentekakis, I.V., & Vayenas, C.G., *ibid*, 1994, **264**, 1563-6
15. Inui, T., *et al*, *Appl. Catal. A: General*, 1993, **94**, 31-44
16. McGhee, W.D., Riley, D.P., Christ, M.E., & Christ, K.M., *Organometallics*, **12**, 1429-33 ∎

Questions

1. What is believed to be the cause of the buildup of carbon dioxide in the atmosphere?

2. What is an advantage of using pigments derived from CO_2 rather than titanium dioxide?

3. List three factors that make CO_2 a good candidate for use in chemical processes.

Answers begin on page 161.

11 *Seventy years after Albert Einstein predicted it, scientists have proved the existence of a new state of matter. By cooling rubidium atoms to the lowest temperature ever achieved, scientists formed a Bose-Einstein condensate, a dense cluster of about 2,000 atoms all in the same state. In effect, the atoms condensed into a "superatom" that behaves as a single entity, according to research leaders. Hundreds of physicists over the decades have attempted to create a Bose-Einstein condensate but failed to reach the extreme low temperatures required. Researchers at the University of Colorado did it by using six laser beams to slow the movement of the atoms.*

Physicists Create New State of Matter

Gary Taubes

By cooling a crowd of atoms to within a hair of absolute zero, researchers have made the crowd behave as one, opening a new arena for physics.

Eric Cornell says that for the first few days after he and his colleagues introduced a new state of matter to this planet, this achievement didn't really sink in. "I really felt kind of numb," says Cornell, a physicist at the National Institute of Standards and Technology in Boulder, Colorado. "It wasn't until my third night afterward that I didn't sleep all night thinking, 'Oh my god, this really happened.'"

What had happened, on the morning of 5 June, was that Cornell and his colleague Carl Wieman, a University of Colorado physicist, had managed to create what is known as a Bose-Einstein condensate (BEC), a gas so dense and so cold—at 180 nanokelvin (billionths of a degree above absolute zero)—that the atoms come to a near standstill. As they slow to a stop, the quantum-mechanical waves that describe each atom spread out and merge, until the entire gas is locked in the same quantum state. In effect, the atoms lose their separate identities and become one. A BEC is to matter what a laser beam is to light: a coherent state in which the usually microscopic laws of quantum mechanics govern the behavior of a macroscopic system.

The prospect of seeing these shadowy laws in action on a large scale had spurred a 15-year quest to create this uncanny material, by snaring and cooling atoms in cages of magnetic force or light. Cornell and Wieman's success, confirmed when they photographed a tiny, durable knot of rubidium atoms at the center of their trap, finally opens the way to realizing that promise.

The achievement also puts to rest some nagging doubts. As nature kept foiling evermore-ingenious schemes for attaining the temperatures and densities of BEC, says Cornell, "people wondered if there was just some reason that Bose condensates were just not meant to be." Even the most promising cooling strategy of recent years, for example, was stymied by a small leak of ultracold atoms from the very center of the trap. By finding a way to plug that leak, Wieman and Cornell, along with Michael Anderson, a postdoc, and graduate students Jason Ensher and Michael Matthews, have put those doubts to rest. "It's a spectacular discovery," says Dan Kleppner, an atomic physicist at the Massachusetts Institute of Technology (MIT) who is a veteran of the quest. "It takes your breath away. This first demonstration is absolutely clear and convincing. It's almost like in a textbook."

The goal was set 70 years ago, when Einstein, building on work by the Indian physicist Satyendra Nath Bose, first predicted this new state of matter. What properties it might have, or even what it would look like, neither Einstein nor any of his successors could say for sure. But the prospect of studying it has been so alluring that ever since the technologies of atom-trapping and -cooling became powerful enough to make BEC seem feasible, a

Reprinted with permission from *Science*, vol. 269, July 14, 1995, pp. 152–53.
Copyright 1995 American Association for the Advancement of Science.

friendly competition has been on to create it (*Science*, 8 July 1994, p. 184).

Each group has its favored technique and even its favored material—atoms of either hydrogen or so-called alkali metals such as sodium and rubidium—but virtually every experiment has built on the successes of its rivals. And virtually everyone in the field has passed at one time or another through Kleppner's laboratory or that of his former student David Pritchard, also at MIT.

Wieman, for instance, had worked with Kleppner for 2 years as an undergraduate at MIT before going off to Stanford University and then joining the University of Colorado in 1984. Kleppner's own strategy for approaching BEC is to refrigerate hydrogen atoms, then trap them in a magnetic field and let the hotter ones fly out of the trap, leaving colder ones behind—a technique known as evaporative cooling. As early as last year this strategy seemed to have put Kleppner, working with his MIT colleague Tom Greytak, in the lead in the race to achieve BEC.

In Colorado, however, Wieman took up a competitive technique, known s laser cooling, which works best with alkali atoms. Far cheaper and simpler than Kleppner's refrigeration apparatus, later cooling works by bombarding atoms with photons. The laser is tuned to a frequency slightly too low for the atoms to absorb when they are at a standstill. But when the atoms are moving toward the laser, the frequency of the light as seen by the atoms is Doppler-shifted upward and the photons hit home, slowing the atoms and hence cooling them.

But although laser cooling can chill atoms to less than a millionth of a degree above absolute zero, a temperature even colder than Kleppner and Greytak had achieved for hydrogen, it could not match the densities they had reached. Many-fold colder temperatures or higher densities—or both—would be needed for BEC. Wieman, however, hit on a possible solution: If he could laser-cool rubidium atoms first and then evaporatively cool them as Kleppner and Greytak had done with hydrogen, he might reach the threshold. When Cornell—a former Pritchard student—joined him in 1990, they set out to make it happen.

To pull off this marriage of techniques, they transferred atoms that had already been laser-cooled to within a few tens of millionths of a kelvin to a magnetic trap. Providing the magnetic axes of the atoms are all lined up in the same direction, the trap can snare them within a "quadrupole" magnetic field, which is strong at the edges of the trap but falls to a spot of zero field at the very center. The hottest, fastest atoms are allowed to escape, leaving colder atoms behind. The strategy cools the atoms by another factor of 5; at the same time, it boosts their density as they cluster at the center.

That was the good news. At a meeting in Anaheim, California, a year ago, however, the Colorado group and others pursuing the hybrid strategy—which also included an MIT group led by Wolfgang Ketterle—faced the bad news: the spot of zero field at the center of the trap. The coldest atoms do indeed cluster at the center as predicted, but because there's no magnetic field there to keep them aligned, they leak out, leaving the researchers four orders of magnitude short of the density-temperature threshold of BEC.

Ketterle proposed one fix: Aim a laser beam at the zero point to repel any atoms that approached it. Ketterle's laser-plugged trap worked, but not as well, or at least not as quickly, as the idea Cornell came up with, which even Ketterle calls "a real gem."

Cornell simply added a second magnetic field to his existing trap, one that would swing the zero point around in a circle—which is one reason why he and Wieman call the result a TOP trap. "The TOP trap," says Wieman, "simply takes this zero field point and moves it away from the center and spins it around. Now what you've got is this orbit of death. As long as the atoms are cold enough to stay in the center and not get out to the orbit of death, they stay there forever and you can keep cooling them down."

By the end of May, the researchers were confident they could acieve the temperature and density needed to create a BEC, but they hadn't figured out how to see it if they did. The Bose condensate would consist of a few thousand atoms in a ball 10 microns across, too small to allow them to see whether the atoms' velocities had dropped to the levels of BEC—"All we'd see is a little smudge," says Wieman.

The solution was what they called ballistic expansion: They would open up the trap, leaving the atoms free to fly apart. Says Cornell: "We wait for a while, and the cloud gets a lot bigger, and then we take a picture of the cloud" using a laser. The structure of this expanding cloud, he says, reveals the velocity distribution in the original cloud before the trap was opened. Hotter atoms should have spread out, but at the center, the density of the atoms should rise steeply. These, he says, are the relic of the BEC that existed until the trap was opened.

That was the theory, anyway. On 5 June, the predicted density peak appeared on the experimenters' video screens. "It was so close to what we have been telling people that it ought to look like that we were initially kind of suspicious," says Cornell. Now his doubts have vanished. Asked whether he and his colleagues could be wrong, Cornell says simply, "I hope not," which he quickly amends to "No. The data are pretty clean. We're not averaging data for 300 hours, getting a [weak] effect, and just happy-talking ourselves into seeing what we want to see."

Indeed, the very first images seem to have cleared up some of the theoretical speculation about what a BEC might be like. For starters, says Wieman, "it takes a couple seconds for it to form, and there's some interesting physics behind that." But one of the most intriguing questions about BEC remains to be answered: What does it look like? Light waves should interact very differently with atoms whose own wave functions have merged than with ordinary matter, but theorists' predictions about the result have been all over the map, with some opting for transparency, some for inky blackness, and some for a silvery sheen. "We have finessed that whole issue by letting the stuff expand out before we ever look at it," says Wieman. But now, he says, it will be very easy to shine a laser on it before it expands and see what happens.

And that's just the beginning of the scrutiny they plan for their prize, he adds. "A thousand atoms [of BEC] is a big chunk, and we'll be able to make a lot more." Then "we will have a whole bunch of [other] knobs we can turn" to probe the new material.

Cornell and Wieman won't be the only researchers doing so. Perhaps half a dozen experiments are still on the verge of creating a BEC, and other groups are likely to join them, inspired by the Colorado group's success—and by the low price of admission. Costing perhaps $50,000 for hardware, plus several months of labor, Wieman and Cornell's apparatus was breathtakingly cheap by modern physics standards. "We worked hard to pursue this using techniques that were cheap and simple," says Wieman, "so if it actually worked it would be opening up the field." ■

Questions

1. To what temperature were atoms of rubidium cooled?

2. What instruments were used to cool the rubidium atoms?

3. What scientist first predicted this new state of matter?

Answers begin on page 161.

12 *You may not be aware of the latest product in the bottled-water industry: bottled water specifically intended for infants. Bottled-water sales have soared in recent years as consumers look for water that has a fresher taste. A part of the trend is probably based on fears that some municipal water is contaminated or at least not very healthy. However, the usefulness of special water for infants is viewed in this report as unnecessary and in some cases dangerous.*

Tapping the Market for Bottled Baby Water

As the bottled-water industry continues to enjoy sparkling success, with sales soaring some 160 percent since 1984, it's no wonder that baby-food manufacturers have started dipping into the waterworks as well. Beech-Nut rolled out Spring Water touted as "bottled with your baby's development in mind" in 1991, and Gerber is currently "testing the waters" in Chicago and several other cities. Even some major drugstore chains, such as Osco, offer bottled water specifically intended for infants.

But the usefulness of special water for babies, which can cost as much as $1.19 a liter, is questionable at best. According to the American Academy of Pediatrics, babies get all the water they need from breast milk or formula during the first six months or so of life. The only time an infant might need a little extra water is during extremely hot weather. But even then, most pediatricians recommend no more than about 4 ounces of water during a 24-hour period—between feedings.

Pushing infants in the nursing stage to drink excessive amounts of water or using it in place of breast milk or formula is not only unnecessary; it can be dangerous. That's because when babies consume too much water, the fluid is absorbed into the bloodstream in amounts large enough to dilute the blood. As a result, the blood's concentration of sodium may fall dangerously low, causing symptoms that include irritability and sleepiness. Known as hyponatremia or water intoxication, the condition can even cause seizures if left untreated.

Unfortunately, many parents are not aware that excess water may be harmful to infants, so that bottled-water labels with pictures of infants may only add to the confusion. Consider that in the fall of 1993, two eight-week old infants in Wisconsin landed in an emergency room suffering with seizures caused by water intoxication. One of the infants' mothers reportedly gave her child bottled water because she thought the label indicated that the water was an appropriate feeding supplement. The other mother, whose baby was suffering from a cold, gave him bottled water because she thought it might help relieve his symptoms. In still other cases, babies who ended up with water intoxication had been fed bottled water as a treatment for diarrhea. Apparently, parents chose the water because it can be easily confused with liquids known as electrolyte solutions, which are typically recommended for diarrhea sufferers and are often stocked right next to bottles of water. These products, which should only be used under a pediatrician's supervision, contain not just water but also the electrolytes sodium and potassium as well as sugar.

In response to the cases of confusion that have trickled in, the Food and Drug Administration has asked baby-water bottlers to state clearly on their labels that the products are *not* electrolyte solutions and should not be used as a treatment for diarrhea or vomiting infants. We did some checking of the market ourselves, however, and found that while Gerber and Beech-Nut both note on the labels that their baby waters are not electrolyte replacements, at least one other product carried no such warning. What's more, of the 51 drugstores around the country we contacted that sell bottled baby water, all but five stock it in the infant-care section of the store, next to or near electrolyte solutions. Many experts think that baby water, which can be mistaken for an inexpensive version of an electrolyte solution, should be stocked in the bottled water sections of drugstores and supermarkets instead.

Reprinted with permission, *Tufts University Diet & Nutrition Letter*. Subscription information: 1-800-274-7581.

The Fluoride Factor

Yet another potential point of confusion for parents is that the various brands of bottled baby water contain fluoride. According to the Beech-Nut label, for instance, "Beech-Nut Spring Water is bottled with your baby's development in mind by starting with spring water and adding fluoride. Fluoride is important for your baby's developing teeth, even before they're visible."

While it's true that fluoride is indeed important for preventing cavities in youngsters, parents need not buy special baby water. For one thing, the American Academy of Pediatric Dentistry recently recommended that infants living in areas where tap water does not contain adequate amounts of fluoride need not be given the cavity-preventing mineral in any form until the age of six months. At six months, fluoride supplementation should start, but pediatricians rarely, if ever, suggest that parents go out and buy fluoridated bottled water for the purposes of diluting juice, mixing with cereals for babies, or using for drinking water. A much cheaper alternative is prescribing fluoride supplements, which can cost as little as $4 a month—a bargain compared with, say, the $1.20 or so per liter of Beech-Nut Spring Water.

Parents should also note that more is not better when it comes to fluoride. If a child swallows fluoridated water in addition to taking fluoride supplements, the excess fluoride could lead to a condition called fluorosis—a staining and/or mottling of tooth enamel that's harmless but undesirable cosmetically.

To be sure, there's nothing wrong with using bottled water to make formula, or offering it to children once they start eating solids. But unless the municipal or well water that flows through your home faucet is unfit to drink for some reason, there's no need to. Even parents who simply feel more comfortable using bottled rather than tap water need not buy a brand that's marked specifically for babies. ■

Questions

1. Why don't babies need extra water?

2. What is hyponatremia?

3. What is the difference between baby water and electrolyte solutions?

Answers begin on page 161.

13 *The future health of the U.S. economy is dependent on a solution to the serious energy shortages this country faces. Domestic oil reserves are projected to be exhausted in 10 to 15 years, depending on consumption patterns. The United States now imports about one-half of its oil at an annual cost of approximately $65 billion. Increased U.S. dependence on foreign oil has serious implications for national security as well as the economy. With these concerns, a number of scientists are very interested in methane hydrate: enormous underground deposits of methane locked in water. Methane hydrate, an icelike material that exists where pressures are high and temperatures are below 20 °C, represents a tremendous potential energy resource.*

Gas Hydrates Eyed as Future Energy Source

Ron Dagani

An icelike material that occurs in underground deposits all over the world and is composed largely of water and methane may turn out to be an unconventional energy source for the future—if scientists can find a way to tap it.

Global deposits of this material, known as methane (or gas) hydrate, are estimated to contain twice as much carbon as all other fossil fuels on Earth. But tapping this enormous potential energy resource may not be as easy as drilling for oil or natural gas.

Nevertheless, scientists and engineers are now beginning to develop strategies to exploit methane hydrate. And some of their ideas were aired in late February in Atlanta at the annual meeting of the American Association for the Advancement of Science.

Methane hydrate is a molecule of methane encaged inside a spherical cluster of water molecules held together by hydrogen bonding. It forms from natural gas and water in sediments and permafrost, where temperatures are low (below 20 °C) and pressures are high (typically 20 bar or more). Gas hydrates can contain small amounts of other gases such as ethane or hydrogen sulfide. The hydrates typically are found beneath the ocean floor on continental margins and in region north of the Arctic Circle (such as the North Slope of Alaska and parts of Canada and Siberia).

The problem with methane hydrate is it is a solid that is not amenable to conventional gas and oil recovery techniques, says Gerald D. Holder, professor of chemical and petroleum engineering at the University of Pittsburgh. Holder and his colleagues have identified two potential recovery techniques for gas hydrates. One involves injecting a warmer fluid, such as seawater, into the hydrate deposits, causing the hydrate to dissociate into water and gas, which can then be recovered conventionally. Another technique involves reducing the pressure in the hydrate deposit by drilling into it to remove free gas that has been trapped by the deposit. This also leads to the breakdown of the hydrate.

An analysis of each technique indicates that pressure reduction may be superior. But the best choice for any deposit will depend on its location and the nature of the surrounding environment, Holder says.

Since known reserves of conventional natural gas are adequate to fill near-term needs, most companies aren't willing to invest in the development of new, unproven technology for recovering gas from methane hydrates, Holder says. But the Japanese, who are energy hungry, aren't waiting for the rest of the world. They are about to embark on a five-year plan that will culminate in exploratory drilling to see whether methane can be recovered from hydrates.

The environmental effects of such drilling are uncertain. Some scientists have expressed concern that the release of methane from hydrates into the atmosphere could affect Earth's climate since methane is an even more potent

Reprinted with permission from *Chemical & Engineering News*, vol. 73, no. 10, March 6, 1995, p. 40. Copyright 1995 American Chemical Society.

greenhouse gas that carbon dioxide. Methane release from hydrate deposits is believed to occur naturally, but large-scale drilling could significantly increase the amount of gas entering the atmosphere.

Many other uncertainties cloud the picture. But the key one is that scientists don't really know whether they will be able to extract methane from hydrate deposits. "If it is resource," comments one geologist, "it's one that's way out there in the future." ∎

Questions

1. What is methane hydrate?

2. What type of bonding forces hold the hydrate together?

3. Why can't conventional gas- and oil-recovery techniques be used to recover methane hydrate?

Answers begin on page 161.

14 *The precise control of the outcome of chemical reactions is a goal long sought by chemists and chemical engineers. To have such mastery over starting materials would enable researchers to fashion exotic substances, new drugs, and other useful compounds in good yield and at lower expense. Recent advances in laser technology are making progress toward this goal. With their ability to deliver energy to tiny targets, lasers are being used to make or break chemical bonds and manipulate the outcome of chemical reactions. If lasers can enhance drug synthesis by yielding the desired compound free of unwanted side products, some medications could become safer and cheaper.*

Dances with Molecules: Controlling Chemical Reactions with Laser Light

Richard Lipkin

Even in the world of large, clumsy objects, a mere flash of light can alter the course of matter.

Consider this scene. A thousand cars descend on a football stadium on a Sunday afternoon. In one hideously long line, drivers queue up to park. A single light with alternating arrows controls the traffic flow. When a green arrow flashes left, cars roll into lot A. When another arrow points right, they cruise into lot B. After an hour or so, the lots contain about 500 parked vehicles apiece.

But suppose someone wants to change the rate at which the two lots fill or the distribution of autos between them—putting, say, 800 vehicles in one lot and 200 in the other. What simple maneuver would do the trick?

The answer, in this particular case, proves to be rather simple. Just change the timing of the traffic light.

In a crude sense, this analogy gives a feeling for the way chemists want to use lasers to control the rate and direction of chemical reactions.

Given a set of molecules that can combine in two possible ways, scientists wonder how they can most effectively use a laser to prod the chemical reaction one way or the other—that is, to direct the molecules as if they were cars rolling past a forked intersection.

"Historically, most chemistry has been done by mixing elements together and heating them," says Richard N. Zare, a chemist at Stanford University. Just blend the ingredients, add a little heat and pressure, and see how a compound cooks up. "But the trouble with heating something is that the energy shows up in the molecules as random motion," he adds. "The energy breaks old chemical bonds, makes new ones, and overcomes barriers to transitions."

The tantalizing question is how to perform more efficient chemistry. Or, as Zare puts it, "Is there a better way to run a chemical reaction?"

Lasers offer intriguing possibilities. With their ability to deliver small, uniform bundles of energy to tiny targets, lasers have in recent years spurred many chemists to rethink how to trigger reactions.

In theory, photons of just the right energy can drive atoms into excited states, make or break chemical bonds, or become absorbed by some molecules while being deflected by others. In practice, though, such precise control has proved elusive in the laboratory.

Only recently have several teams of chemists begun to show how to put theory into practice by controlling some simple molecular reactions with lasers.

Robert J. Gordon, a chemist at the University of Illinois at Chicago, and his colleagues recently achieved "coherent phase control" of hydrogen disulfide molecules by firing ultraviolet lasers of different wavelengths at them. The two beams agitate the molecules in different ways. By varying the phase difference—the relative locations of the crests and valleys of the waves—of the two lasers, the researchers can control how the molecules break apart.

Beginning with a molecule that contains two hydrogen atoms and one sulfur atom, they can control its rate of ionization. Their report, which appeared in the April 8 JOURNAL OF CHEMICAL PHYSICS, constitutes the first observation of a successfully directed chemical reaction for a molecule with more than two atoms.

Recently, in a similar experiment using molecules of hydrogen iodide, Gordon's group succeeded in varying the amounts of two different reaction products merely by altering the phase difference between the two lasers.

"This work is very exciting," says David J. Tannor, a chemist at the University of Notre Dame in South Bend, Ind. "It's important because Gordon's group is successfully using the phase properties of laser light to manipulate the outcome of a photochemical reaction."

By finely adjusting the relative phases of the two lasers, the researchers can push the molecules into specific high-energy states.

"The laser expands the atoms," says Gordon. "An electron goes into a high orbit so that it's ready to be ejected. The molecule can then either release an electron or break a bond."

This expansion sets the stage for a specific reaction to occur.

"If you have two different ways of producing a molecular product," says Gordon, "then the phase difference between the light of each laser can be used to regulate the speed of the reaction. If more than one outcome is possible, the beams can steer the reaction in a desired direction."

"Ultimately, we want to be able to select specific chemical bonds and break them with a laser pulse," Gordon adds. "That would permit us to increase or decrease one reaction product versus another."

Taking a different approach to stimulating molecules with lasers, Kent R. Wilson, a chemist at the University of California, San Diego, and his colleagues are "shaping" laser pulses rather than creating interference with two beams.

The key idea behind Wilson's method falls into the same general camp as Gordon's in that both groups employ coherent light to direct the outcome of a reaction. The difference, however, shows up in the way the researchers deliver the light energy. Rather than adjust the interplay of two distinct light beams, Wilson's apparatus emits a series of ultrashort bursts of energy, or femtosecond pulses, each lasting a few millionths of a billionth of a second.

Each pulse comes as a carefully sculpted wave of energy, Wilson says, and in some cases bears a distinctive chirp, or frequency signature. By tinkering with the shape of a laser's light waves, the scientists can deliver energy to a molecule with tremendous accuracy and efficiency, striking it in exactly the right way to trigger a specific reaction.

Reporting in an upcoming PHYSICAL REVIEW LETTERS, Wilson's team describes how its laser delivers tailored "vibrational wave packets" of energy to molecules of iodine, pumping the atoms into excited states. Using theory to figure out how best to shape the laser pulse, the team says the experimental results coincide well with prediction.

"We did not just excite matter with light and observe the results as we changed the properties of the pulse," the scientists state. "We chose a goal for our material system, predicted the best possible field to reach that goal, and then attempted to create that field in the laboratory."

Indeed, excitement has arisen in the chemical community not because these experimental results prompt an entirely new set of ideas, but because they confirm the suspicion that what once appeared possible but improbable can now be done.

"Ever since the development of the laser, the quest to use light to control the future of matter has been one of the Holy Grails of chemistry," Wilson says.

"Can we develop general ways to use lasers to specifically manipulate the quantum behavior of atoms and molecules?" Wilson and his coworkers ask in ACCOUNTS OF CHEMICAL RESEARCH, 1995, issue 3. "Can we develop novel synthetic methods to produce exotic new molecules, states of molecules, or even molecular devices, such as programmable optics or nanomachines?"

Such radical thinking permeates the field of laser chemistry. Nearly a decade has passed since theoreticians first proposed schemes to achieve laser-mediated molecular control. In 1986, chemists Paul Brumer of the University of Toronto and Moshe Shapiro of Israel's Weizmann Institute wrote of "actively" controlling matter with light by way of quantum mechanical interference. Shortly thereafter, David Tannor and chemist Stuart A. Rice of the University of Chicago proposed an alternative method to achieve the same effect with short bursts of laser light.

These two theoretical paradigms for laser chemistry dominated the experimental world, shaping the goals of many laboratories. As it turns out, both theoretical visions have led to important results. Gordon's group, for example, evidences the Brumer-Shapiro scenario, while Wilson's group fleshes out the Tannor-Rice model.

Taking yet another tack toward laser control of molecular behavior, Paul B. Corkum and

Emmanuel Dupont, physicists at Canada's National Research Council in Ottawa, have shown how laser light can change the direction of electrical current flows in a semiconductor.

In the May 1 PHYSICAL REVIEW LETTERS, the two scientists describe how to combine two laser beams in such a way that their phase difference causes optically stimulated electrons to move right or left in a semiconducting material. Moreover, the effect is not random, the scientists explain. They can reliably control the amount of current produced, as well as its direction of flow.

In all three approaches, the researchers have had to overcome formidable technical obstacles in order to control electrons with laser light. Though the concept of laser chemistry beckoned for decades, "experimental examples were limited," says Zare. "But the recent results of Gordon, Wilson, and Corkum are encouraging. Hopefully, they will stimulate more experiments."

An alternative approach to laser-controlled chemistry derives from the work of physical chemist Herschel A. Rabitz of Princeton University. Rabitz and his colleagues have used the mathematics of control theory to improve the efficiency of photochemical reactions. Unlike big solid objects, molecules constantly move about, rotating and vibrating according to their level of energy. In Rabitz's approach, sculpted laser pulses alter the internal motions of molecules: To vibrate a molecule in a particular way that will trigger a reaction, you need to know what kind of light pulse would get that molecule twitching.

"You might say that the goal of our research is to use light to get molecules to dance to our tunes," Rabitz says.

So what might come of all this laser chemistry? Are its goals realistic?

Brumer suggests that the first applications will appear in the pharmaceutical industry. Many molecules come in both right- and left-handed forms. In some cases, one of these forms constitutes a beneficial drug, while the other causes harm.

Drug companies spend great amounts of time and money separating one group of molecules from the other. If lasers could enhance drug synthesis, however, yielding large quantities of desired compounds free of unwanted side products, some medications could become safer and cheaper.

Likewise, laser-controlled semiconductors could promote new optical switches, potentially improving computers and communications systems.

"In 1986, theory was way ahead of experiment in this field," says Tannor. "People thought laser chemistry was impossible to do." Now, he says, "the situation has reversed. Experiment is ahead of theory."

An explosion in laser technology in the past few years has given former pipe dreams some solidity. "We're still far away from fully understanding the complete internal motions of molecules," Tannor says. "But we're on the verge of being able to interrogate reactions very thoroughly, which will give us very detailed knowledge of how molecules move."

Such knowledge might lead not only to atom-by-atom construction of molecules and materials, but also perhaps to improved ways to gather solar energy, Tannor says. Earth's most effective solar energy systems—green plants—can transform the sun's rays into chemical energy with up to 95 percent efficiency. "It's possible," says Tannor, "that the knowledge we gain from light-controlled chemistry could someday lead to synthetic light-harvesting systems."

Wilson speculates that the quest to control matter with light may lead some chemists to pursue "late-night dreams." And yet, he adds, "like all dreams, they may not come to pass."

"We hope that the process of our dreaming, and the progress of our search, may lead us to outcomes that are of value," Wilson muses, "even if they are not our original goals." ∎

Questions

1. Why will being able to select and break specific chemical bonds with laser light be useful?

2. How can lasers be used to affect semiconductors?

3. Why might pharmaceutical companies be interested in laser chemistry?

Answers begin on page 161.

15 *Scientists and world leaders have been concerned for a decade about the ozone depletion taking place over the Antarctic. Now it is reported that this past winter brought record ozone loss to the northern (Arctic) polar regions. A form of oxygen, ozone (O3) is a colorless gas that at the level of the earth's surface is a major contributor to smog. However, the ozone layer in the stratosphere, centered some 30–35 km above the earth, plays a very essential role. It protects life at the surface of the earth by absorbing 95–99 percent of the sunlight in the 200–300 nm wavelength range (the ultraviolet range). Light in this range is especially damaging to living forms and would have drastic effects on our planet. Evidence is very strong that the widespread use of chlorofluorocarbons (CFCs) and closely related bromine and chlorine compounds known as* halons *is depleting the ozone layer.*

Complexities of Ozone Loss Continue to Challenge Scientists

Pamela S. Zurer

Severe Arctic depletion verified, but intricacies of polar stratospheric clouds, midlatitude loss still puzzle researchers.

This past winter brought record ozone loss to the Arctic polar regions, scientists at an international conference confirmed last month. While reluctant to call the severe ozone destruction a "hole" like the one that develops each year over Antarctica, European researchers presented a convincing case that the depletion in the far north resulted from the same halogen-catalyzed chemistry that triggers the Antarctic phenomenon.

But few other issues addressed at the weeklong International Conference on Ozone in the Lower Stratosphere, held in Halkidiki, Greece, could be resolved with as much certainty. Atmospheric scientists are still struggling to grasp the exact nature of the polar stratospheric clouds that are so crucial to the fate of ozone in the polar regions. The dynamics of the stratosphere remain imperfectly understood. And questions persist about the mechanism of the gradual ozone-thinning trend over the midlatitude regions of North America, Asia, and Europe, where most of the world's people live.

Without a doubt, a broad outline of the complex behavior of stratospheric ozone is in focus. And—as the research presented at the conference attests—scientists are working diligently to fill in the details. Until the remaining uncertainties are clarified, however, the ability to quantitatively predict the condition of the ozone layer will remain elusive, especially as chlorine levels are expected to peak over the next few years and then slowly decline over several decades.

The goal of the recent meeting was to bring together U.S. and European scientists to share their results. "It is vital [that] information is exchanged as widely as possible within the world scientific community," said University of Cambridge chemist John L. Pyle, one of the organizers. Designed to provoke discussion, the conference featured a handful of invited talks and more than 200 poster presentations. Its sponsors included the European Union (EU), the World Meteorological Organization (WMO), the National Aeronautics & Space Administration (NASA), and the National Oceanic & Atmospheric Administration (NOAA).

For some of the 300 scientists from 40 nations who attended, the meeting marked a return trip to the lush Halkidiki Peninsula in northern Greece. Eleven years ago they had gathered to discuss ozone depletion at the very same beach resort.

Back then, depletion of stratospheric ozone by chlorine and bromine from man-made chemicals was only a hypothesis. But the discovery of the Antarctic ozone hole the following year brought theory to life, kindling a surge of research and political activity. In the decade since, the atmospheric science community has greatly improved its understanding of stratospheric ozone through a combination of field observations, laboratory experiments, and computer modeling. Meanwhile, the nations of the world have joined in the Montreal Protocol on Substances That Deplete the Ozone Layer to phase out production of chlorofluorocarbons (CFCs) and halons, the major sources of ozone-depleting halo- gen compounds in the stratosphere.

On the minds of many researchers as they arrived at the conference was the question of Arctic ozone depletion during the winter just past. For some years, evidence has accumulated that the Arctic winter stratosphere is loaded with the same destructive chlorine species believed to cause the Antarctic ozone hole. Only the normally milder northern winters have prevented massive ozone loss.

Atmospheric scientists have been predicting that a prolonged cold Arctic winter—more like those usually experienced in Antarctica—could unleash severe ozone destruction. Arctic stratospheric temperatures hit new lows in the early months of 1995, and preliminary reports indicated that ozone had decreased dramatically (C&EN, April 10, page 8).

Indeed, WMO's network of ground-based ozone-monitoring instruments observed record low ozone over a huge part of the Northern Hemisphere, averaging 20% less than the long-term mean, Rumen D. Bojkov told the conference. Bojkov, special adviser to the WMO secretary general, said the period of extreme low ozone began in January and extended through March.

"In some areas over Siberia, the deficiency was as much as 40%," he said. "It's only because of normally high ozone in that region that we do not have what you would call an ozone hole."

Measurements from ground-based instruments alone are not enough to prove that Arctic ozone has been destroyed chemically, however. Changes in ozone could also result from the constant motion of the atmosphere transporting air with different ozone concentrations from one place to another.

"It's difficult to tell if the variations are chemical or dynamical," mused NASA atmospheric scientist James F. Gleason. "Could an ozone high over Alaska have been balancing the low over Siberia?" Unfortunately, there have been no daily high-resolution satellite maps of global ozone to help answer that question since NASA's Total Ozone Mapping Spectrometer (TOMS) aboard Russia's Meteor-3 satellite failed in December.

Even if TOMS data were available, it would be extraordinarily difficult to untangle chemical destruction of ozone from dynamical fluctuations. The exception is the evolution of the Antarctic ozone hole each September. The stratosphere over Antarctica in winter is isolated by a circle of strong winds called the polar vortex, so ozone concentrations there are normally at a stable minimum before the ozone hole begins to form. The dramatic ozone depletion that occurs as the sun rises in early spring is unmistakable.

The Arctic polar vortex, in contrast, is usually much weaker, and ozone concentrations in the north polar regions are constantly changing. They normally increase in late winter and early spring, bolstered by waves of ozone-rich air from the tropics. So even if ozone amounts hold steady or increase somewhat in the Arctic, ozone still may have been destroyed: The concentrations may be significantly less than they would have been had there been no chemical depletion. Researchers have resorted to ingenious methods of calculating Arctic ozone loss indirectly—for example, by studying the changing ratio of the amount of ozone to the amount of the relatively inert "tracer" gas nitrous oxide in a given parcel of air.

This past winter, however, the Second European Stratospheric Arctic and Midlatitude Experiment (SESAME) generated a wealth of data that is helping to overcome the complications posed by ozone's tremendous natural variability. The 1994–95 EU-sponsored research campaign employed aircraft, balloons, and ground-based instruments to study the stratosphere. European scientists eagerly presented their latest findings at the Halkidiki conference—all of which point to widespread chemical destruction of Arctic ozone.

One elegant experiment used coordinated balloon launches to measure ozone loss directly during the Arctic winter. The trick is to measure the amount of ozone in a particular parcel of the air, track the air mass as it travels around the Arctic polar vortex, and then measure its ozone content again some days later. The approach was described by physicist Markus Rex, a doctoral student at Alfred Wegener Institute for Polar & Marine Research (AWI) in Potsdam, Germany.

In a previous experiment, Rex and his coworkers used wind and temperature data from the European Center for Medium-Range Weather Forecasts to identify which of some 1,200 ozone sondes—small balloons carrying electrochemical ozone sensors—launched during the winter of 1991–92 intercepted the same air mass at two different times. From the matches they identified, they estimate over 30% of the ozone at 20 km within the Arctic polar vortex was destroyed during January and February of 1992 [*Nature*, **375,** 131 (1995); C&EN, May 15, page 28].

Rather than again rely on chance matches as they had in 1991–92, the researchers coordinated releases of more than 1,000 ozonesondes from 35 stations during this past winter's SESAME campaign. After launching a balloon, the scientists used meteorological data to forecast its path. "As the air parcel approached another station, we asked that station to launch a second sonde," Rex said.

The scientists observed that ozone was decreasing throughout January, February, and March 1995. And they found the decline in a given air parcel was proportional to the time it had spent in sunlight, consistent with photochemical ozone depletion catalyzed by halogens.

Rex and his coworkers calculate that ozone was being lost at a rate of about 2% per day at the end of January and even faster by mid-March, when the sun was flooding a wider area. Those depletion rates are as fast as those within the Antarctic ozone hole.

"The chemical ozone loss coincides with, and slightly lags, the occurrence of temperatures low enough for polar stratospheric clouds," Rex said. Such clouds provide surfaces for reactions that convert chlorine compounds from relatively inert forms to reactive species that can chew up ozone in sunlight.

The results of other SESAME experiments add to the conclusion that the Arctic suffered severe ozone loss last winter. For example:
• Temperatures in the Arctic stratosphere in winter 1994–95 reached the lowest observed during the past 30 years, reported Barbara Naujokat of the Free University of Berlin's Meteorological Institute. The north polar vortex was unusually strong and stable, breaking up only at the end of April.
• Profiles of the vertical distribution of ozone within the polar vortex revealed as much as half of the ozone missing at certain altitudes compared with earlier years, reported AWI's Peter von der Gathen.
• An instrument carried by balloon into the stratosphere above Kiruna, Sweden, recorded high concentrations of chlorine monoxide (ClO) in February, said Darin W. Toohey, assistant professor of earth systems science at the University of California, Irvine. Chlorine monoxide, the "smoking gun" of ozone depletion, forms when chlorine atoms attack ozone. Simultaneous ozone measurements showed substantial amounts missing.

These and many other findings coalesce into an "incredibly consistent" picture of Arctic ozone depletion, said Cambridge chemist Neil Harris. But was there actually an Arctic ozone hole this year?

"I wouldn't call it a hole," said Lucien Froidevaux of California Institute of Technology's Jet Propulsion Laboratory (JPL).

"At most we've got half a hole," said Cambridge's Pyle.

"We don't have to be so hesitant," said NOAA research chemist Susan Solomon. "The data show ozone was not simply not delivered but actually removed. This is exciting confirmation of substantial Arctic ozone depletion. It's never going to look exactly like an Antarctic ozone hole, but so what?"

Whatever one chooses to call what happened in the Arctic earlier this year—one wag suggested "Arctic ozone dent"—key questions remain unanswered. Will the severe depletion return in subsequent winters, intensify, or increase in area? How are the dramatic ozone losses in the polar regions affecting stratospheric ozone over the rest of the globe?

"Yes, I believe there have been statistically significant changes in Arctic winter ozone that we can observe," said NOAA chemist David Fahey. "But where do we go from here? Can we predict future changes?"

One issue hampering atmospheric scientists' ability to make qualitative predictions of future ozone changes is the difficulty in understanding the exact nature of polar stratospheric clouds (PSCs). For several years after their critical importance to the Antarctic ozone hole was discovered, researchers thought they had a good grasp of the situation. But reality has turned out not to be so neat, said Thomas Peter of Max Planck Institute for Chemistry, Mainz, Germany.

It is the presence of PSCs that makes ozone in the polar regions so much more vulnerable than it is in more temperate regions. The total amount of chlorine and bromine compounds is roughly uniform throughout the stratosphere. The halogens are carried there by CFCs and halons, which break down when exposed to intense ultraviolet light in the upper stratosphere.

In most seasons and regions, the halogen atoms are tied up in so-called reservoir molecules that do not react with ozone—hydrogen chloride and chlorine nitrate ($ClONO_2$), for example. However, PSCs—which condense in the frigid cold of the stratospheric polar vortices—provide heterogenous surfaces for reactions that convert the reservoir species to more reactive ones.

The most important reaction is between the two chlorine reservoirs:

$$ClONO_2 + HCl \rightarrow Cl_2 + HNO_3$$

The molecular chlorine produced flies off into the gas phase, where it is photolyzed easily by even weak sunlight to give chlorine radicals—active chlorine—that can catalyze ozone destruction.

Equally important is the fate of the nitric acid (HNO_3) produced by the heterogeneous chemistry. It remains within the PSCs, effectively sequestering the nitrogen family of compounds that would otherwise react with active chlorine to reform chlorine nitrate. That process, called denitrification, allows the photochemical chain reactions that destroy ozone to run efficiently for a long time without termination.

Laboratory and field experiments have confirmed the importance of heterogenous reactions in the winter polar stratosphere. What is at question now is the actual composition of the PSCs, how they nucleate and grow, their surface area, and their chemical reactivity. All those factors affect the interconversion of halogen compounds between their active and inactive forms, and thus the rate and amount of ozone depletion.

Peter described the "happy period" in the late 1980s when scientists were confident they understood just what PSCs are. Type I PSCs were thought to be crystals of nitric acid trihydrate (NAT) that condensed on small sulfate aerosol particles once temperatures cooled below about 195 K. Frozen water ice (type II) appears once stratospheric temperatures plunge lower than about 187 K, which generally happens only in Antarctica.

Now it appears type I PSCs are not so simple. "Say 'bye bye' to the notion PSCs must be solid," Peter said. "They can be liquid. Both types of particles are up there."

Margaret A. Tolbert, associate professor of chemistry and biochemistry at the University of Colorado, Boulder, explained that "Everybody assumed type I PSCs were NAT, which condenses about 195 K. But observations show nothing actually condenses until about 193 K. That doesn't prove the particles aren't NAT. But they may instead be supercooled ternary solutions" of water, sulfuric acid, and nitric acid.

Such supercooled solutions could develop from small sulfate aerosol particles. (The stratosphere contains a permanent veil of sulfate aerosol droplets, which form when sulfur dioxide from volcanic eruptions and carbonyl sulfide emitted by living creatures are oxidized to sulfuric acid.) As the stratosphere cools in winter, the sulfate aerosols could take up water and nitric acid, growing larger but remaining liquid.

Whether type I PSCs are solid or liquid can make a significant difference, said NOAA chemist A.J. Ravishankara. He noted that chemistry could occur not just on the surface of supercooled solutions, but also in the interior of the droplet, which he likened to a little beaker. That implies scientists would have to consider not just the surface area of the particles but their volume in calculating the rates of chemical processes involving PSCs.

Furthermore, he said, the supercooled solutions may persist over a larger temperature range than solid NAT particles. That would allow transformation of halogens to their active forms to take place over a wider temperature range than previously recognized.

The implications of chemical processing taking place on and in supercooled ternary solutions extend beyond the polar regions, where PSCs appear, to more temperate regions of the globe. Although changes in ozone in the midlatitude stratosphere are nowhere near as dramatic as in the polar regions, they are real and substantial. The latest United Nations Environment Program study, "Scientific Assessment of Ozone Depletion," concludes that ozone in the mid-northern regions, for example, has been decreasing at a rate of about 4% per decade since 1979.

Atmospheric scientists have been struggling to quantitatively explain that decrease, which is predominantly in the lower stratosphere. Chlorine and bromine radicals are clearly implicated, but in the absence of PSCs they destroy ozone most voraciously at much higher altitudes where there simply isn't all that much ozone to begin with. Modelers have been plugging every known ozone destruction cycle into their calculations but still have not been able to account for all of the ozone loss observed below about 20 km.

Roderic L. Jones, of Cambridge's chemistry department, noted that chemistry involving supercooled ternary solutions could have significant effects on ozone trends at midlatitudes. "Look at areas in the Northern Hemisphere that are exposed to temperatures just above the NAT [condensation] point, about 197 or 199 K," he said. "It's a big area, extending as far south as 50° N."

Supercooled solutions may turn out to play an important role in explaining what's going on at midlatitudes. But JPL modeler Ross J. Salawitch said he thinks the problem may arise from the way scientists approximate the dynamics of the stratosphere in their models.

Two-dimensional models treat ozone as if it diffused out uniformly from the tropical stratosphere where it is produced. "The real world is more like the 'tropical pipe'" paradigm, Salawitch said, in which air rises high into the stratosphere in the tropics and moves downward again at higher latitudes, like a sort of fountain.

"The issue of whether the global diffusion or the tropical pipe model better represents reality is not just an academic discussion," Salawitch said. "The question of a midlatitude deficit may disappear when the models handle dynamics better."

Answers to some of the issues that continue to bedevil atmospheric researchers may become clear as more data accumulate. Participants in the SESAME campaign have barely had time to think about what they observed. And NASA has just begun its new three-year Stratospheric Transport of Atmospheric Tracers mission.

But the stratospheric ozone layer may reveal even further complexities in the coming years. As

Christos S. Zerefos, local organizer of the conference from the laboratory of atmospheric physics at Aristotle University of Thessaloníki, Greece, pointed out in his closing remarks, the ozone layer will only begin to recover if the nations of the world continue to comply with the Montreal protocol. If not, atmospheric scientists may have even greater puzzles to contend with. ∎

Questions

1. What experimental method is used to track ozone depletion in the Arctic?

2. What are the major sources of ozone-depleting halogen compounds in the stratosphere?

3. What makes the ozone in the polar regions more vulnerable to destruction?

Answers begin on page 161.

16 *Three prescription drugs promising a significant breakthrough in treating acid indigestion are being reintroduced as a powerful new generation of over-the-counter (OTC) stomach acid blockers. Much of the discomfort of overeating results when the stomach overreacts and pours out too much gastric hydrochloric acid (HCl) to help digest the food. This excess supply of stomach acid leads to acid indigestion and heartburn. The common remedy is to take an antacid, like Tums or Mylanta, which typically contains a carbonate or hydroxide. These neutralize the stomach acid that has already been produced, but the effect only lasts about two hours. The three new OTC drugs—Zantac, Tagamet, and Pepcid—promise a significant breakthrough by blocking the production of stomach acid and providing all-day relief.*

Drug Giants Ready to Enter Stomach-Medicine Battle

Steve Sakson

New York—Call it the battle of the bellyache.

Three of the world's most popular prescription drugs—ulcer medicines called Zantac, Tagamet and Pepcid—are being reintroduced as a powerful new generation of over-the-counter stomach acid blockers.

They have the potential to revolutionize the $1 billion-per-year market for antacids, which hasn't seen a major innovation since the last century. Their fight for supremacy promises to be one of the drug industry's most tenacious.

The drugs promise a significant breakthrough—they block production of stomach acid, providing all-day relief from indigestion, heartburn and sour stomach.

That's welcome news for the millions of Americans who have spent a day popping Tums or slugging Mylanta after a hefty serving of spicy meatballs.

By late this year or early next, industry analysts predict the trio of new products will be common sights on television, magazines and billboards as manufacturers gear up advertising campaigns expected to cost at least $100 million each.

"I think we will be blitzed," said Mariola Haggar of C.J. Lawrence-Deutsche Bank in New York.

And for good reason.

"We're talking about three gladiators here," said David Saks of Gruntal & Co. "It would be a major disappointment if they, as a combined group, didn't double the size of the gastrointestinal therapy market" within three years, he said.

Drug companies often get their latest over-the-counter products from the ranks of their prescription medicines and these entries are the biggest ever to make the switch.

Tagamet, introduced by Britain's SmithKline Beecham PLC in 1977, became the first drug in history to sell $1 billion a year. Zantac is now the best selling drug in the world, with annual sales of about $3.6 billion.

The drugs are the best-known members of a family called H2 blockers. As prescriptions, these pills have treated hundreds of millions of patients with gastric ulcers or other gastrointestinal diseases.

Antacids, typically made from chemicals like calcium carbonate or magnesium hydroxide, neutralize stomach acid that has already been produced. While they go to work in a few minutes, they last only about two hours—the time they stay in the stomach.

H2 blockers must be absorbed into the bloodstream, so the onset of relief is longer, perhaps up to an hour. Once there, however, they slow down or stop acid from being produced for six to 12 hours, doctors say.

They act by blocking proteins called H2 receptors in the stomach's acid-producing cells from triggering a chemical reaction that causes the cells to secrete acid. The effect is similar to the antihistamines people commonly take for a stuffy nose to cut the production of mucus.

While Tagamet, Pepcid and Zantac vary widely in potency, the

Copyright 1995 Associated Press. Reprinted with permission.

non-prescription dosages are being structured so all three will have about the same effectiveness, doctors say. Therefore, their success depends largely on the timing of their introduction and how they're promoted.

On that score, Pepcid has an obvious lead. The drug's creator, Merck & Co., in partnership with its New Jersey neighbor Johnson & Johnson, persuaded the Food and Drug Administration to approve non-prescription Pepcid AC (for acid controller) in April and rushed it to market by early June.

J&J-Merck is doing saturation advertising, hoping to ingrain Pepcid's name in the minds of consumers before the others even hit the market.

SmithKline Beecham got approval for Tagamet HB (for heartburn) in June and expects to begin selling it in the early fall.

Zantac, made by Britain's Glaxo Wellcome PLC, hasn't gotten FDA approval yet and won't be introduced until late this year at the earliest, analysts say.

However, that delay may be irrelevant once Zantac arrives. As a prescription drug, Zantac outsells the others by more than 4-to-1, a market dominance guaranteed to make its non-prescription version a major player.

Nonetheless, Pepcid has other advantages besides timing. J&J-Merck did extra studies on humans and got permission to sell Pepcid for a new use—preventing indigestion, not just curing it.

"Every consumer can relate to certain foods that always cause them heartburn—chili, Mexican or spicy food," said Merck spokesman Gary Lachow. "We believe people can prudently eat meals they once were unable to, by taking a Pepcid AC an hour prior," he said.

Although doctors say Zantac and Tagamet will do the same, Glaxo and SmithKline never asked that their drugs be used that way.

It's a decision that SmithKline vice president Jack Ziegler, for one, said he doesn't regret. "Very few consumers can actually predict when they're going to get heartburn," said Ziegler.

Pepcid's other advantage concerns safety—or at least the perception of it.

On rare occasions, some H2 blockers have interfered with other medicines. The FDA required SmithKline to include a warning on Tagamet HB's label telling people who take three other drugs to consult their doctors before taking it. The drugs are warfarin, a blood thinner used mostly by heart patients; theophylline, an asthma drug; and phenytoin, an epilepsy drug whose most popular brand name is Dilantin.

Zantac may also be required to make such warnings, but the FDA didn't require them for Pepcid because the risks are considered smaller.

"I'm not making a claim that Pepcid is safer, but we do have the tremendous benefit of not having any warnings vis a vis Tagamet and I assume we'll have the same benefit vis a vis Zantac," said Dr. Patrick Ciccone, a J&J-Merck vice President.

Doctors, however, dismiss any safety concerns as extremely minor. Robert Fisher, chairman of the gastroenterology section at the Temple University School of Medicine in Philadelphia, said Tagamet in particular has gotten a "bum rap."

"The truth is those effects haven't bothered a lot of people. It's one of the safest medications ever introduced. (Pepcid) may be a little safer."

Despite their blockbuster potential, analysts point out that profits from these drugs probably won't come for two years. The biggest reason is the tremendous start-up and promotional costs.

In addition, Stephen Gerber, an analyst for Oppenheimer & Co., expects some consumers may get turned off by the drugs' slow-acting nature compared to current over-the-counter remedies.

If the drugs do sell well, they will steal sales of antacids made by the same companies. SmithKline, for instance, makes Tums and J&J-Merck makes Mylanta.

In coming years, there will be other competitors, including cheaper generic versions and perhaps other brand names like Eli Lilly & Co.'s H2 blocker, Axid.

The drugs will also be expensive. A box of 12 Pepcid sells for $5.50 at a discount drug chain in New York City, while 75 Tums go for $2.39.

Analyst Haggar doubts however that consumers will be bothered by sticker shock when they see the drugs on store shelves.

"Pepcid costs about 45 cents a pill, but it may provide relief for a whole day. That's less than a cup of coffee," she said. ■

Questions

1. How do antacids work?

2. Why is the onset of action for H2 blockers slower than that for antacids?

3. What is an advantage of H2 blockers over antacids?

Answers begin on page 161.

17 *Sufficient supplies of energy are essential for supporting our present standard of living, planning for the future, and even maintaining national security. With a major landfill site near many communities in the United States, landfill gas may play an important role as an alternative energy source. Landfill gas is a combustible mixture of gases produced by microbial action on the organic material present in landfills. It consists primarily of methane, carbon dioxide, and hydrogen. Because methane forms explosive mixtures with air, controlling the release of landfill gas is important. Great Britain has been particularly active in addressing the serious environmental problems associated with landfill gas while harnessing the methane for energy use.*

Having the Last Gas

Nigel P. Freestone, Paul S. Phillips, and Ray Hall

Alternative energy sources are frequently the butt of jokes, but with vast areas of land being claimed for waste burial, landfill gas is making a serious challenge as a potentially important commodity.

Millions of visitors to the 1985 International Garden Festival in Liverpool were probably unaware that the landscape gardens they had come to enjoy were on top of approximately 10m t of domestic rubbish. Nor were they aware that this rubbish was generating a methane-rich gas that had necessitated installing a special extraction system to remove it from the site. At first this gas—landfill gas—was simply flared (burnt), but once it was realized that extraction would be needed for at least 15 years it was decided to use the gas to generate electricity. This development is a good example of the potential, especially in the developed world, for using landfill gas produced in refuse tips as an additional and versatile energy source.

An Organic Bioreactor

Landfill gas is the mixture of gases produced by microbiological activity on putrescible/biodegradable or other similar organic material deposited in landfill sites. Indigenous anaerobic bacteria digest the decaying organic matter that comprises a large proportion of the contents of domestic dustbins. When large amounts of refuse accumulate, the landfill site acts as a bioreactor in which microorganisms produce a biogas typically composed of approximately 35–40 per cent carbon dioxide and 60–65 per cent methane.

It is theoretically possible for each tonne of waste in a landfill site to provide 400 m^3 of gas. Calculations of how long the gas will continue to be generated are necessarily only estimates, but 10–15 years is accepted as a reasonable, perhaps conservative, estimate of the time for high levels of gas to accumulate, and biogas will continue to be produced for 50–100 years. Over 10 years 1 t of domestic rubbish can produce more than 100 times its own volume in biogas, equivalent to 10,000 MJ of energy. Landfill gas could provide 2.5m t of coal equivalent each year from existing sites in the UK, using available technology. By carefully controlling the methods for landfilling and consolidating new sites throughout the disposal operation, however, site operators will soon be able to optimise gas production and could provide the UK with up to 6.5m t of coal equivalent each year, assuming that there will be continued landfilling of organic wastes. The possible expansion of schemes for using such wastes in compost production or for anaerobically digesting it in bioreactors would have a significant impact on gas generation rates at landfill sites receiving different waste mixtures. However, the organic content at landfill sites containing less putrescible wastes may be more than made up by the proposed use of sewage sludge.

Generating large quantities of landfill gas is a fairly recent phenomenon: until the early 1970s, landfill sites for municipal wastes were small, producing correspondingly small amounts of gas, which simply leaked to the atmosphere. No doubt the local people registered the smell, and occasionally there were fires, but the gas was not considered a serious hazard. In 1974 municipal waste disposal became the responsibility of county authorities (in England), and larger landfill sites began to be used. Since

Reprinted with permission from *Chemistry in Britain*, January 1994.

the end of World War II the biodegradable content of domestic waste has increased by 15 per cent or more, according to some estimates. Although it is difficult to gauge this value accurately, it has certainly increased in line with the overall quantities of waste per capita, primarily resulting from social changes such as the decline of domestic composting and the increasing use of retail packaging.

Microbial Activity

Landfill gas composition varies considerably with time.
• Initial landfill site conditions favour aerobic digestion, quickly exhausting all the available oxygen and causing all the aerobic bacteria to die.
• Conditions will now be moist with little oxygen, ideal for anaerobic bacteria.

To promote bacterial growth, landfill sites should be alkaline. The majority of UK landfill sites do not knowingly restrict incoming waste to between pH 4 and 7. However, the initial rapid breakdown of the putrescible components of household waste produces an acidic leachate. Many sites taking household waste also take 'inert' demolition and construction wastes and these act as a buffer, preventing acidic leachate from forming if intimately mixed with household wastes. Traditionally, anaerobic digestion has been considered as a two stage process, a non-methanogenic stage followed by a methanogenic stage. The non-methanogenic stage has also been referred to as the acid-forming stage because volatile fatty acids are the principal products. It is now known that this first stage may itself include as many as three steps:
1. The first, involving the hydrolysis of organic waste—fats, proteins and polysaccharides—produces long chain fatty acids, glycerol, short chain peptides, amino acids, monosaccharides and disaccharides.
2. The second step (acid formation) involves forming a range of relatively low relative molecular mass materials, including hydrogen, methanoic acid and ethanoic acid, other fatty acids, ketones and alcohols. It is now established that only methanoic and ethanoic acids can be used as substrates by the methanogenic bacteria.
3. In the third step, substances other than methanoic acid and ethanoic acid are converted by hydrogen producing acetogenic (OHPA) bacteria. Some bacteria are able to undertake both the first and second steps, producing hydrogen, methanoic acid and ethanoic acid; therefore the third step is no longer needed.

Anaerobic digestion is sensitive to fairly low concentrations of toxic pollutants such as heavy metals and chlorinated organics. If the balance of the process is upset it is likely that the methanogenic organisms will be inhibited first, resulting in a build-up of intermediate compounds immediatley before methane formation.

As well as carbon dioxide and methane, other gases such as carbon monoxide, hydrogen and some air, or at least nitrogen with other trace gases are also present in landfill gas. To date, over 140 different compounds have been identified in landfill gas, ranging from simple hydrogen sulphide to complex halogenated organosulphur compounds, depending on the types of waste digested.

Operating Problems

The importance of controlling the release of landfill gas was highlighted in 1986 when gas from a refuse tip in the Derbyshire village of Loscoe—where a depth of over 70 m of rubbish had accumulated over 10 years—ignited, demolishing a bungalow and injuring the occupants; 350 houses nearby had to be evacuated. A series of other explosions and incidents the following year, including at least 14 other occasions when houses threatened by gas migrating from tips into foundations were evacuated, led Her Majesty's Inspectorate of Pollution to ask all local authorities to assess the safety of their landfill sites. The results suggested that, of the landfill sites receiving putrescible wastes, over 60 per cent that were then active and 75 per cent that were closed in the preceding 10 years were (or had been) generating enough gas to cause problems in the future unless properly managed.

At all sites receiving biodegradable material there is bacteriological breakdown and consequently gas, heat and moisture are produced, causing slow settling of the landfill ground and seepage of gas and leachate. Leachate is a dark, highly polluted liquid that forms when rainwater or water from surrounding streams permeates the site. Typically containing organic materials, metal chlorides and sulphates, it has a high chemical oxygen demand—a measure of the amount of potassium dichromate needed to oxidise any reducing materials in the sample—and its pH ranges from 5.8 to 8.5.

In the past there have been serious environmental problems because of leachates moving off site, for example via permeable rock; nowadays almost every new site built is waterproofed by being lined with clay or other impermeable materials. Leachate still escapes from some older sites; however, in these so-called dilute and attenuate sites the rocks or other materials around the site can neutralise harmful species in the leachate and disperse it, minimising pollution. The major pollution risk is the movement of leachate into groundwater via rock fissures, which is extremely difficult to control.

Containment sites have collection systems to pump any leachate that builds up to an aerobic treatment plant where oxygen is pumped through it to reduce its polluting effects. Dilute and attenuate sites may also build up levels of leachate if incoming liquid exceeds the amount dispersed.

Producing landfill gas by digesting solid waste creates a pressure at the site, leading to the diffusion of gas out of the site and into the surrounding environment. Landfill gas is both explosive and an asphyxiant in air, so its egress from a site needs to be carefully monitored and controlled. The gas will always take the path of least resistance and hence its movement is affected by site operating practice. Normal practice is the so-called 'onion skin' method, whereby waste is laid and compacted in thin layers, a small area at a time.

Good working practices and site preparation can minimise unwanted gas migration. Lining the site with low permeability materials—either a sufficient thickness of clay or an artificial polymeric liner—should stop egress from the site via any artificial tunnels or rock fissures. Site operators must also take other precautions to ensure that gas does not collect in any man-made void space such as cellars or manholes around the site. The more gas produced, the greater the pressure within the site and the more gas that will be forced out, either through horizontal fissures or vertically. The rate of gas output depends on atmospheric pressure; the greater the atmospheric pressure, the smaller the pressure difference and hence the slower the evolution of landfill gas and *vice versa*. Gas generation is not only affected by atmospheric pressure but also by other weather conditions that might seal the ground, for example, a hard frost or, more problematically, snow. These conditions might lead to a dangerous buildup of gas pressure, giving a higher risk of explosions. In addition, an increase in pressure within the site will cause more gas to dissolve in leachate, which in turn may migrate off site. As the leachate moves away from a site, or as it moves up to ground level, the pressure decreases, releasing the dissolved gases.

Collection Systems

At sites from which landfill gas is being used, one of three types of collecting system (depending on conditions at the site) is installed for transferring the gas directly to where it is needed:

Monitoring boreholes. The basic design of monitoring boreholes is a wide hole bored into the waste, which is then filled with aggregate around a perforated plastic pipe. The top of the borehole is sealed with a layer of cement, sometimes below a layer of bentonite clay (which is very impermeable to gas), so that the gas percolates through the pipe. These boreholes are placed strategically around the whole site, allowing gas migration to be monitored.

Gas wells. Gas wells are similar to boreholes but with an extraction system attached. Two types of extraction systems are currently used: passive and non-passive. In passive systems the gas wells are arranged in a grid formation, typically 20 m apart, with the grid extending to 10–15 m from the edge of the landfill, to optimise gas collection. The efficiency of the well depends on the permeability of the surrounding material, so each well produces different quantities of gas. In a non-passive system, where the gas is sucked out for use, the wells can be further apart, but generally not a greater distance than 40 m.

Gas pumping system. With the installation of the gas collecting system, there is usually a flare stack, possibly with an electrical generating engine. These are connected to the collection wells by thermoplastic pipes of diameters typically in the range 10–15 cm, depending on the flow rate, with some sort of pump connected in line. Gas pumping needs to be carefully controlled to ensure that the pumping rate does not exceed gas production and thereby draw in air.

Gas Exploitation

By the late 1970s landfill gas was being extracted from several sites, and by the early 1980s such sites were increasing in size. A landfill site in Essex, for example, operated by the then Greater London Council, was accepting 250,000 t pa of domestic waste from east London and depositing it in a 40 m deep pit over an area of 250,000 m². The site was harnessed for methane production in 1982 and the gas—3500 m³ per hour—sold to Thames Board at Purfleet.

At current energy prices, the UK landfill gas resource is estimated to be worth over £125m a year. In 1992, 33 landfill gas schemes were active in the UK, with a total energy saving equivalent to 143,000 t of coal a year.

In 1990 the UK government initiated the Non Fossil Fuel Obligation (NFFO)—a scheme encouraging the generation of electricity from non-fossil sources, which includes renewable sources and landfill gas, in England and Wales. Under this obligation, the privatised electricity generating companies pay a levy from the revenue of all electricity generated from fossil sources. The proceeds of the levy are used to improve the financial viability of schemes from renewable and other non-fossil sources by paying higher tariffs for the electricity generated. Twenty-five landfill gas schemes were accepted in the first part of the scheme, with a total installed capacity of 25 MW; of these 16 are new projects, with an installed capacity of 15 MW. The introduction of the NFFO has given a boost to landfill gas exploitation and it is estimated that by the end of the century around 150–175 MW of electricity will be generated from landfill gas in the UK.

Landfill sites are a significant source of two major greenhouse gases, carbon dioxide and methane; and it is estimated that emissions from landfills account for about 21 per cent of total UK methane emissions. This is the third most important methane source in the UK, behind agriculture (33 per cent) and coal mining (29 per cent). The effect of methane as a greenhouse gas is variously estimated as being up to 27 times that of carbon dioxide, therefore any gas control or utilisation schemes that result in burning the methane will reduce the damaging effects of emissions from landfills. Gas utilisation will have the additional benefit of displacing power generation from fossil fuel sources.

Of the 242 landfill gas schemes that are now operating worldwide, 55 per cent generate electricity—a total capacity in excess of 200 MW. In the UK, private electricity producers can now sell their output to electricity generating companies, and larger schemes produce enough electricity to support the needs of a small town. Current research indicates that their output might be boosted by using sewage sludge, which would increase the organic content. And, in the UK, the privatisation of the water authorities together with stricter controls on dumping at sea may well open up new possibilities for co-operation over sewage sludge disposal in landfills, with a resulting increase in gas generation rates.

Extraction Problems

Explosive risk. Methane forms explosive mixtures with air at concentrations between five and 15 per cent. This is obviously different in an extraction system where the atmosphere should be almost entirely landfill gas (there is always a small proportion of oxygen).

Smells. Examining a typical landfill gas site reveals a variety of trace substances such as H_2S, organosulphur compounds, and various hydrocarbons. Although these compounds may be present in minute amounts (*ca* 0.001 per cent), their aroma is obvious to anyone who has smelt landfill gas. This smell is a problem even when an extraction system is present—if the gas is not burnt properly by a flare stack or generating engine, at a high enough temperature, incomplete combustion occurs, causing noxious smells in the exhaust gases. The smell changes depending on incineration temperatures because of the differing 'dwell' times of the gas in the combustion area.

Building near sites. HM UK Inspectorate of Pollution stresses that no buildings should be erected on landfill sites. Under the Town and Country Planning General Development Order Regulations of 1988, any planning application referring to an area with 250 m of a landfill site must be assessed for the risk of landfill penetration. However, completed landfill sites could be used for farming or for environmental/recreational purposes as well as for producing landfill gas.

Further Reading

1. *Landfill gas; waste management paper no 27.* Department of the Environment, London: HMSO, 1992.
2. *Monitoring of landfill gas.* Northampton: Institute of Wastes Management, 1990.
3. *Landfill sites: development control.* Joint circular DOE 17/89; Welsh Office 38/89.
4. *Landfilling wastes: A technical memorandum for the disposal of wastes on landfill sites: waste management paper no 26.* London: HMSO, 1986. ■

Questions

1. Describe the composition of landfill gas.

2. Why is the first stage of anaerobic digestion referred to as the acid-forming stage?

3. What risk does landfill gas pose during extraction?

Answers begin on page 161.

18 *The chemistry of filtration is extremely important in industries ranging from brewing beer to refining petroleum. Zeolites, a class of very porous materials, play a key role in the filter industry. Zeolite filters have traditionally been made from porous crystals of inorganic materials such as aluminum and silicon. Recently, chemists have made progress in developing organic zeolites. Researchers at the University of Illinois and the University of Michigan have reported making organic zeolites out of large molecules with preformed pores. The ability to vary the size of these pores to match the needs of the filtration process holds great promise for future studies.*

Making Molecular Filters More Reactive

Robert F. Service

Organic chemists are on the verge of something big in the manufacture of molecular strainers called zeolites. These nanoscopic filters, used widely by industry, have traditionally been made from porous crystals of inorganic materials such as silicon and aluminum. But zeolites made of organic materials could open up an exciting new range of applications in areas including drug manufacture. Before this can happen, however, organic chemists must be able to control the size of the pores in the zeolite—and so far they've only been able to make extremely small pores.

At last week's American Chemical Society (ACS) conference in Washington, D.C., however, a team led by Jeffrey Moore, an organic chemist at the University of Illinois (UI) at Urbana-Champaign, and Stephen Lee, an inorganic solid state chemist at the University of Michigan (UM) at Ann Arbor, reported making organic zeolites out of large molecules with preformed pores. When these ring-shaped molecules stacked up, they formed channels as large as 17 angstroms in diameter—more than twice the width of the earlier organic record. The discovery opens the door to making organic zeolites with even larger pores, simply by increasing the size of the holes in the prefab building blocks.

"It's very interesting," says Galen Stucky, a professor of chemistry and materials at the University of California, Santa Barbara. "Historically, inorganics have been used [to make zeolites] because their cost is cheap and you can make them in bulk quantities. But the thing you have with organics is the potential to make small variations [in the building blocks] to control the chemistry of the framework."

The zeolite construction crew started to make their prefab pores using molecules that look like the spokes and hubs of a child's Tinker Toy set. The spokes were rigid acetylene molecules and the hubs, ring-shaped benzenes. UM synthetic chemists Jinshan Zhang and Ziyan Wu connected these chemical Tinker Toys one by one into a hexagon, with benzene hubs at the corners, joined by acetylene spokes. Finally, Zhang and Wu added hydroxyl groups to the outside of each hub; these groups can form hydrogen bonds to join one hexagonal pore to a neighboring one.

When the researchers placed these hexagonal pores in a solution of ethanol and methanol, the hydrogen bonds drew neighboring hexagons together to form a two-dimensional honeycomb pattern. Intermolecular forces caused additional layers to stack up, creating a solid crystal with channels 8 angstroms in diameter. Using similar starting materials, the group also fashioned a more complex 3D building block that led to their second organic zeolite with 17 angstrom channels.

With their new organic zeolites in hand, the UI/UM team plans to join reactive molecules to the inside of the pores, giving their strainers highly specific binding properties that can't be built into inorganic zeolites. One such possibility, says Moore, is adding molecules such as

Reprinted with permission from *Science*, vol. 265, September 2, 1994, p. 1363.
Copyright 1994 American Association for the Advancement of Science.

amino acids that preferentially bind to one of a pair of mirror-image, or chiral molecules. Such left- or right-handed molecules are commonly used as drugs, and pharmaceutical manufacturers must separate such nearly identical twins, as in many cases one molecule is therapeutic while its mirror image is either ineffective or toxic. And currently, the most widely used process for separating left- and right-handed versions of the same molecule requires several chemical stages.

The trick in developing such filters, says chemist Thomas Bein of Purdue University in West Lafayette, Indiana, "will be to add functional groups without affecting self-assembly of the crystal." Unlike inorganic zeolites, in which molecules are held together with rugged covalent bonds, organic zeolites are glued with relatively weak intermolecular forces and hydrogen bonds. And the same reactivity that makes the functional groups candidates for placement inside the pores could potentially affect how the molecules interact. The fragility of these bonds also means that organic filters won't replace inorganic zeolites in high-temperature applications, such as breaking apart the large hydrocarbons in crude oil into the smaller ones used in gasoline. But if organic chemists can manage to add these groups, they'll have a field of zeolite applications all to themselves. ■

Questions

1. Why have zeolites traditionally been prepared from inorganic materials?

2. What is an advantage of organic zeolites?

3. Why are organic zeolites relatively unstable at higher temperatures?

Answers begin on page 161.

19 *Chlorine (Cl_2) reacts with most organic compounds, even with the alkanes, which are known for being unreactive. Because of this high reactivity, there are a tremendous number of organic chlorine compounds. Many of these are quite useful substances. Vinyl chloride, for example, is used to make polyvinyl chloride, or PVC, the plastic in water pipes, floor tiles, and synthetic leather. At the same time, a great number of organochlorines tend to be toxic and harmful to the environment. Thus, a deepening controversy surrounds the use of chlorine and its compounds.*

The Chlorine Controversy

Gordon Graff

Some environmentalists are calling for the total elimination of the many chemicals containing chlorine. While that will not likely happen, government and industry could take measures to reduce the use of unsafe compounds.

Chemicals containing chlorine—*all* chemicals containing chlorine—have become increasingly suspect. Amid evidence that some chlorinated pesticides, solvents, plastics, and even byproducts of elemental chlorine from its use in water purification are harming wildlife and perhaps people, environmentalists and legislators are calling for sharp cutbacks in the production, use, and discharge of chlorinated products.

Critics include not only the environmental activist group Greenpeace, which has launched a high-profile campaign to totally eliminate chlorine compounds, but others calling for less extreme though still serious measures. In August 1993 Rep. Bill Richardson (D-N.M.) proposed a bill that would require pulp and paper manufacturers to phase out the use of chlorine in bleaching operations.

Such bleaching is known to produce traces of dioxins and other potent compounds that can induce cancer. The International Joint Commission (IJC), an environmental-policy group organized by the U.S. and Canadian governments that focuses on the Great Lakes region, has recommended banning chlorine and chlorine-containing compounds as industrial feedstocks. Several European organizations—the Oslo and Paris Commissions for the Prevention of Marine Pollution, and International Whaling Commission, the World Wilderness Conference, and the Barcelona Convention on the Mediterranean Sea—have echoed the IJC recommendations. And most recently, in March 1994, the American Public Health Association, a professional society, called on industry to reduce or eliminate chlorinated organic compounds in its products and processes and to introduce "lower-risk" substitutes. The list of charges against chlorine is lengthy. According to critics, chlorine compounds are slowly poisoning the earth, insidiously working their way through the air, groundwater, and food chains, contributing to ozone depletion, destroying wildlife, and causing cancer and infertility in humans as well as a host of other ills. "Overwhelming" numbers of organic compounds containing chlorine—organochlorines—"tend to be toxic and persistent in the environment," says Joe Thornton, Greenpeace's research coordinator. "This doesn't mean that all chlorine compounds behave the same way, but virtually every organochlorine that's ever been tested has been found to cause at least one significant adverse effect."

Chlorine is used in roughly 15,000 products with estimated annual U.S. sales of $71 billion. Many of these products have become integral to industrial society. Polyvinyl chloride (PVC) plastics, for instance, are cheap, strong, and durable enough to find service in everything from house siding and sewer pipes to food wrapping and toothbrushes. The vast majority of commercial pesticides and pharmaceuticals, such as the herbicide atrazine, vitamin supplements, and cardiovascular and central-nervous-system drugs, also contain chlorine or are manufactured using chlorine-based chemistry. And chlorine and chlorine-based compounds are used to disinfect 98 percent of the pub-

Reprinted from *Technology Review*. Copyright 1995, Gordon Graff.

licly supplied drinking water in the United States.

Manufacturers of chlorinated chemicals say that such compounds are not intrinsically environmental culprits. Instead, industry representatives maintain, various chemicals in all classes can be toxic, persistent, and bioaccumulative—that is, can build up in living tissue. Brad Lienhart, managing director of the Chlorine Chemistry Council, a Washington, D.C.-based trade group formed last year to fight chlorine restrictions, cites as an example polyaromatic hydrocarbons, which are produced when hydrocarbons such as oil are burned. Others include compounds of heavy metals like mercury, lead, and cadmium. Meanwhile, he says, "a very high percentage" of chlorine compounds "are neither toxic, persistent, nor bioaccumulative."

The chemical industry supports the system used by the U.S. Environmental Protection Agency (EPA) and its state counterparts today, in which risks are identified and exposure guidelines set for individual chlorine-based chemicals. But critics find this approach unacceptable. Regulatory agencies need "to deal with the whole field of chlorine" at once since "organochlorines are not formed one by one; they're formed in complex mixtures," says Thornton.

In February of this year EPA weighed in for the first time on the subject. While announcing the Clinton administration's plan to reauthorize the Clean Water Act, EPA administrator Carol M. Browner said the agency "will develop a national strategy for substituting, reducing, or prohibiting the use of chlorine and chlorinated compounds." The statement, which actually amounted to a proposal to study the feasibility of reducing some uses of chlorine, satisfied neither side. It predictably rankled an industry concerned that this was the first step toward a chlorine ban, and it dismayed environmentalists who felt that it didn't go far enough.

But it wasn't an unreasonable idea for an agency seeking middle ground in a debate over how to protect humans and wildlife without imposing unacceptable economic and quality-of-life burdens. A growing body of scientific evidence suggests that both EPA and industry should consider taking a variety of measures to reduce the production and use of chlorine compounds.

Persistent When Dangerous
At the root of the controversy over chlorine are properties that have proved a blessing and a bane: the element is extremely reactive and the resulting compounds extremely stable. Strongly negative chlorine atoms tend to snatch electrons away from other atoms, accounting for the element's high reactivity with electron-rich carbon compounds and the diverse panoply of organochlorine compounds that are made possible by such reactions.

The strong bonds that chlorine tends to form with other atoms make some, but not all, industrial chlorine compounds highly resistant to the forces that normally break down natural substances and many nonchlorine chemicals. For example, PCBs—chlorine-containing insulating fluids used in electrical transformers—are still ubiquitous in the environment despite having been banned in 1976 because of links to human cancer.

Because they break down extremely slowly, chlorinated pesticides, solvents, and industrial byproducts such as dioxins may accumulate in lakes, rivers, and streams, ultimately affecting health and reproductive abilities of wildlife and perhaps humans. Researchers have amassed much evidence of specific effects. In the summer of 1991, leading environmental scientists from academia, government, and environmental organizations met to compare notes about their studies of chemical effects on wildlife. They concluded that some organochlorines—including the herbicide atrazine, the insecticide oxychlordane, and the industrial chemical pentachlorophenol—as well as certain nonchlorinated compounds such as the pesticides amitrole, carbaryl, and parathion could disrupt the endocrine systems of both wildlife and humans. That system controls many aspects of development and behavior through the production and release of hormones.

The researchers linked this effect to a host of recently observed ills in fish, birds, and mammals, including thyroid dysfunction, decreased fertility, gross birth deformities, and metabolic and behavioral abnormalities. The scientists also drew an association between these chemicals and sexual abnormalities: masculinization of female fish and birds and feminization of male fish, birds, and mammals. The researchers further connected organochlorines to damage to the immune systems of birds and mammals, leaving them prey to infectious diseases.

Even more worrisome are recent reports of connections between endocrine-disrupting organochlorines and human problems. Scientists including reproductive physiologists Richard M. Sharpe of the British Medical Research Council's Center for Reproductive Biology and Niels Skakkebaek of Denmark's National University Hospital have blamed organochlorines for various reproductive-tract cancers as well as low sperm counts and, more generally, male infertility, both of which seem to be on the rise in industrial societies. In 1993 Mary S. Wolff, an associate professor of environmental and occupational medicine at the Mt. Sinai School of Medicine in New York, showed a positive correlation between breast cancer and blood-serum levels of DDE, a chemical breakdown product of DDT, in a group of 58 women. Wolff and other researchers have cited the ability of certain chlorinated compounds to mimic the action of estrogens, female sex hormones, as a possible cause of this effect. And Joseph Jacobson and Sandra Jacobson, psychologists at Wayne State University in Ohio, claim that pregnant women who eat fish exposed to high levels of PCBs have given birth to abnormally small babies who later exhibit behavior problems and intellectual deficits when they reach school age.

But the case for adverse human health effects form environmental organochlorines is not closed. A study published in 1994 by epidemiologist Nancy Krieger of the Kaiser Foundation Research Institute in Oakland, Calif., found no statistical link between the incidence of breast cancer and DDE or PCB concentrations in a group of 150 women. And a paper published last July in the *British Medical Journal* called into question the reliability of sperm-count data that Sharpe and Skakkebaek depended on.

These inconsistencies and uncertainties do not surprise researchers who doubt that organochlorines that are estrogen mimics are present in the environment in quantities large enough to have any effect on human health. "The average human exposure to estrogens is 99.999 percent from natural sources," says toxicologist Stephen Safe of Texas A & M University. Still, most scientists find the evidence of adverse human health effects from chlorinated organic chemicals alarming enough to warrant further exploration.

As long ago as 1972, the U.S. government first banned a chlorinated compound, the pesticide DDT, based on overwhelming evidence about reproductive failure in birds and other wildlife. Through the remainder of the 1970s the government, suspecting links to cancer, also banned or restricted seven other chlorinated insecticides (dieldrin, kepone, mirex, lindane, aldrin, chlordane, and toxaphene). Because of suspected damage to wildlife, Germany, Italy, and the Netherlands have limited the use of the chlorinated herbicide atrazine, although it is widely used as a broad-spectrum agricultural weed killer in the United States. Many individual cities in Germany and Austria have banned the use of PVC plastics, which release dioxins when burned. And for one group of chlorinated compounds there is worldwide agreement: chlorofluorocarbons (CFCs) are being phased out under the Montreal Protocol, the international agreement designed to reverse destruction of the earth's protective ozone layer.

Toward Faster Action
Given the problems linked to a number of chlorinated compounds and their tendency to persist in the environment, Greenpeace's Thornton maintains that the system EPA uses to evaluate their safety—thorough investigation and regulation of one compound at a time—is far too slow to protect humans and wildlife. Consider how the agency assesses the risk level associated with a pesticide. Under present policies, any company seeking to register or, as periodically required, reregister a pesticide must submit voluminous data on its toxic effects. These typically stem from studies in which rats and mice are fed the test chemical from birth, and from test-tube assays such as the Ames test, which evaluates whether the chemical causes genetic mutations in bacterial DNA. In another required study, researchers evaluate the product's effect on the offspring of several generations of rats exposed to the chemical. Finally, registrants must submit data on what happens to fetuses of pregnant rabbits exposed to the chemical. For a pesticide, a field assessment of its environmental fate is also part of the review process.

Such registration proceedings typically take years. Beginning in 1989, for example, EPA started to reregister 800 active pesticide ingredients under tougher guidelines. As of this past April, it had completed the process for only 60 compounds. Companies have been allowed to sell all the products during this period.

Like EPA's pesticide program, its regulation of pollutants discharged by industry into waterways is based on setting limits for individual chemicals—again, a bureaucratically complicated process. The agency now limits the discharge of 126 high-risk "priority pollutants," including chlorinated organic chemicals such as methylene chloride, typically used as an industrial solvent; nonchlorinated organics such as toluene, used as a solvent or for synthesizing other chemicals; and heavy metals such as cadmium, used as, for example, a heat stabilizer in plastics. EPA bases the discharge limits for each chemical on a "best available technology" study conducted for each industry that discharges the compound, and revises the studies every few years.

Unable to speed such evaluations by considering all chlorine compounds at once, the agency now prefers a middle course between industry and environmentalists: an interagency government task force to examine the environmental and health impacts of chlorine and chlorinated compounds and the availability, efficacy, and safety of substitutes. The proposed study would focus on four main uses of chlorine: water disinfection, solvents, polyvinyl chloride and other plastics, and pulp and paper manufacturing. EPA says it will then "develop a plan for any appropriate actions"—a result that could take several years.

Organizations desiring faster action could lobby to amend the Federal Insecticide, Fungicide and Rodenticide Act—which gives EPA authority to regulate pesticides—to allow the agency to restrict or even ban groups of chemicals, such as those in the chlorine family, if many members have known risk factors such as persistence and bioaccumulation. Such a change would require an escape clause for individual chemicals that do not pose significant danger.

Obtaining laws that give EPA broader authority to regulate whole classes of pesticides wouldn't be easy, however. Such proposals have surfaced many times on Capitol Hill, but, says Charles M. Benbrook, an environmental consultant in Washington, D.C., "nothing has ever been passed" because the issue "is so controversial" and unpalatable to industry. EPA's proposed study on chlorine-based chemicals could help proponents of a chemical-class amendment, however, by encouraging a public debate on the costs and benefits of chlorine technology and appropriate courses of action.

EPA could also make its regulation of chlorine compounds more effective by revising its general approach to risk assessment. Under its present policies, the agency estimates a chemical's ability to cause cancer—which may lead to severe restrictions on use or a ban on export—by assuming that a high-dose response can be extrapolated to very small doses. This approach often overstates a chemical's risks because the dose-response curve may not be linear. In fact, as biochemist Bruce Ames of the University of California at Berkeley has suggested, the risk from a chemical toxic to animals at very high doses may drop to virtually zero at the trace levels at which the chemical is likely to be encountered in the environment.

EPA is now proposing changes to its cancer-testing guidelines that would, in effect, allow some "threshold" factors to be considered when evaluating cancer risks of pesticides submitted for registration or reregistration. Such a policy might reassign some chlorinated organics and other chemicals now labeled carcinogens into non-carcinogen or low-level-carcinogen categories.

The proposal would enable EPA toxicologists to consider the mechanisms underlying carcinogenicity rather than simply extrapolate from gross effects in laboratory animals. For example, under the new guidelines researchers might evaluate how a chemical interacts with the endocrine system to produce tumors, in the process finding that the compound produces tumors only when that system is malfunctioning. Richard Hill, science adviser in the Office for Prevention within the EPA branch that deals with pesticides and toxic substances, says that by emphasizing biological processes, the proposed studies would make it easier to evaluate the risks of chemicals tested on animals for humans.

Such a procedure would still require exhaustive animal tests of low doses of thousands of chemicals and is not as systematic as it could be. Most European environmental agencies shave adopted a strategy EPA might do well to consider: screening compounds early in the review process to see if they harm cultured, cellular DNA. Positive results are almost always a red flag for the potential to induce cancer. Technicians then conduct animal tests of compounds testing positive and assume that those proving carcinogenic at high doses have similar effects at low doses. For products that do not harm cellular DNA, scientists conduct *both* low- and high-dose tests in animals to see whether threshold effects exist.

Certainly this testing would take some time. But it would be speedier than the present approach that subjects each chemical, regardless of its likely risks, to a full battery of tests.

A quick-screening assay for the endocrine effects of persistent, bioaccumulative compounds would also be valuable. Ana M. Soto and Carlos Sonnenschein, cell biologists at Tufts University School of Medicine in Boston, have recently developed one such test. It is specific for estrogen mimics—compounds that simulate the effects of the female sex hormones. These mimics, which include both chlorinated and nonchlorinated compounds, are believed responsible for most adverse endocrine effects caused by chemicals in the environment.

In the assay, cultured human breast-cancer cells, which are estrogen-sensitive, are separately exposed to a natural estrogen and the chemical being tested. Technicians then compare the rates of cell proliferation in the two cultures. Test compounds that cause cells to grow as rapidly as natural estrogens are highly estrogenic and therefore could have significant effects. Soto reports that her assay results correlate well with observed estrogenic effects in rodents, with no false positives.

The assay could alert investigators to the need for follow-up trials in animals. Unfortunately, while well-established animal models exist for assessing acute toxicity and cancer, there are few standardized laboratory animal models for gauging suppression of the immune system, reproductive failures, and behavioral changes that specific chemicals may cause. Nor are there reliable ways of extrapolating the findings from such studies to humans.

What Industry Should Do
Meanwhile, even if only to protect themselves from adverse publicity and possible future lawsuits, chemical manufacturers would do well to evaluate their entire slate of chlorinated products to determine which they could replace now.

Precedent for such actions exists. For example, companies began developing substitutes for CFC refrigerants and plastics-processing aids as far back as the 1970s, when it became clear that these compounds posed an environmental threat and might eventually be banned. As a result, the mandated phaseout of CFCs has created minimum economic disruption for manufacturers of these products and their employees.

There is some evidence that industry is already pruning sales of certain marginal pesticides, both chlorinated and nonchlorinated. Many companies declined EPA's invitation to reregister pesticides in the late 1980s, points out Penny Fenner-Crisp, an EPA toxicologist. The lack of interest, she notes, followed EPA's disclosure of tougher—and more expensive—testing requirements than under previous registration procedures. And according to environmental consultant Benbrook, the number of pesticides registered with EPA has dropped by more than half since 1987.

Benbrook notes that the pesticide industry could cut back further. "It would be very easy to reduce by about two-thirds the use of chlorine-based pesticides, with only marginal impact on farmers," he says. He asserts that farmers could lower their chemical reliance by changing their tilling and cultivating practices, although this would require "a little more time." Recognizing the growing concern about chlorine, chemical manufacturers should also consider whether they could lower the overall production of other chlorinated chemicals.

And manufacturers who use chlorine in their production processes could consider shifting to alternative technologies. A case in point is the pulp and paper industry. Even though the standard paper-whitening process of chlorine bleaching is not yet prohibited, as proposed in the Richardson bill now before Congress, some companies have recently begun substituting other bleaching agents. One is chlorine dioxide, which, although chlorine-based, produces far lower dioxin levels than pure chlorine. Other manufacturers have switched to non-chlorine bleaches such as oxygen and hydrogen peroxide, which release no dioxins.

Some technology changes, however, make little sense. The use of chlorine to disinfect water, although it can produce traces of chloroform and other harmful chlorinated organics, is one process that seems unlikely to change for now, since the system is cheaper and widely considered more effective than any alternative. Ozone, the water-disinfection technique cited by Greenpeace as a chlorine alternative, tends to dissipate more rapidly, requiring repeated treatments or follow-up treatments with some chlorine, according to EPA's Science Advisory Board. Moreover, ozone has been shown to form its own harmful products in water, including bromates, which are potent carcinogens. Society will clearly have to determine what tradeoffs it is willing to accept as the chlorine controversy plays out in legislative halls, corporate boardrooms, and the media. ■

Questions

1. What properties of chlorine are at the root of the controversy over its use?

2. Why are the estrogenic effects of chlorine compounds of concern to researchers?

3. What problems does ozone present when substituted for chlorine as a water disinfectant?

Answers begin on page 161.

20 *There is a fine line between the beneficial and harmful effects of alcohol consumption. Alcohol is unquestionably the most abused drug in the United States. It is reported that drunk drivers are involved in more than half of all fatal automobile accidents. Yet, several studies have shown that moderate wine consumption is also associated with lower mortality from coronary heart disease. The mechanism by which wine confers this special benefit may be due to phenolic substances. Phenols, compounds with an -OH attached to an aromatic ring, are widely distributed in fruits and vegetables.*

Wine and Heart Disease

Andrew L. Waterhouse

Phenolic substances may be the compounds responsible for the reduced incidence of coronary heart disease seen in people who regularly consume wine.

Wine has been espoused for centuries as a superior beverage. Louis Pasteur said, "Wine is the most healthful and most hygienic of beverages," and Plato remarked, "No thing more excellent nor more valuable than wine was ever granted mankind by God." Common to these endorsements is the view that wine is as good for the body as it is for the spirit. Modern scientific research supports this perception. David Goldberg at the University of Toronto has recently published an amusing and enlightening review of this subject.[1]

Fifteen years ago, in an ecological epidemiology study, Selwyn St Leger and co-workers showed that there was a population-based association between a reduction in deaths from heart disease and increased wine consumption.[2] More recently, Serge Renaud and Michel de Lorgeril at INSERM in Lyon brought the issue to public attention with a similar study "Wine, alcohol, platelets, and the French paradox for coronary heart disease." They used World Health Organization data to show that dairy fat consumption is highly correlated with coronary heart disease (CHD) mortality. A few French cities, however, had very high fat consumption, yet low CHD mortality rates—thus the "French paradox." When they added wine consumption as a factor that affected CHD mortality, the researchers got a better correlation, with wine being a negative correlate—it appeared to reduce heart disease.[3] Michael Criqui and Brenda Ringel at the University of California, San Diego, subsequently investigated comparable data and came to a similar conclusion: wine was one of the few dietary factors that correlated with reduced CHD mortality.[4] Interestingly, they also showed that fruit consumption correlated with reduced CHD mortality.

There now appears to be no dispute that moderate wine consumption is associated with lower CHD mortality. A related question is whether or not total mortality rates decrease with increased wine consumption, and here there is still some argument. In an ecological study that compared entire populations, Criqui and Ringel[4] showed that total mortality does not decrease as the population's wine alcohol consumption increases. The authors attribute this effect to a compensating increase in mortality from other causes, which offsets the decreasing CHD mortality. However, in prospective studies which distinguish between subjects based on consumption rates, the lowest total mortality occurs with moderate alcohol consumption (1-3 drinks/day), whether from beer, wine or spirits.[5] For heavy drinkers, however, mortality is higher than non-drinkers, especially among women. Other studies agree that the lowest mortality occurs at moderate alcohol consumption levels, including the analysis of health professionals by Eric Rimm and co-workers, at Harvard School of Public Health.[6]

Despite the issues surrounding total mortality, it is clear that moderate alcohol consumption itself reduces CHD mortality. Several mechanisms for this are now recognised, of which the best known is alcohol's ability to alter blood lipid levels by lowering total cholesterol

Reprinted with permission from *Chemistry & Industry*, May 1, 1995.

and raising high density lipoprotein (HDL) levels.[7]

But does wine confer any special benefit, and if so, is there an organic explanation for this effect that is unrelated to alcohol? Criqui and Ringel's ecological data clearly show that wine alcohol consumption (correlation coefficient, r = −0.66) is much more strongly correlated with reduced CHD mortality than total alcohol consumption (r = −0.39). Also when Arthur Klatsky and Mary Ann Armstrong of the Kaiser Permanente Medical Center in Oakland, California, singled out wine drinkers in their prospective study, the drinkers exhibited a lower CHD mortality rate than the other subjects.[8] Thus, in two types of studies, wine appears to have a special benefit. In their US-based study, Klatsky and Armstrong raised a concern that the correlation between wine consumption and reduced CHD mortality may not be due to wine by itself, but perhaps to other lifestyle factors associated with wine consumption. For instance, in the US wine drinking correlates with increased income, which itself is related to reduced mortality rates. On the other hand, Criqui and Ringel's study is not subject to such a bias because the correlation between income and wine consumption does not hold true across the developed nations surveyed.

While future epidemiology studies can be designed to take into account such factors that still raise questions, epidemiological investigations are never capable of determining specific causes for observed effects. Therefore, it is important for chemists and biologists to establish whether or not there is a molecular mechanism by which wine nutrients could affect CHD.

One report has helped to define such an area for investigation. Michaël Hertog and co-workers at the DLO State Institute for Quality Control of Agricultural Products, Wageningen, observed a direct correlation between reduced CHD mortality in elderly men and dietary levels of flavonols, one of the major classes of flavonoid phenolics.[9] The investigators divided the subjects into three groups based on the content of flavonols in their diet. Flavonol consumption was based on their reported diets and the flavonol content of those foods.[10] The best correlation between reduced CHD mortality and specific dietary components was with tea, onions and apples. The best correlation between CHD mortality and chemical constituents of the diet was with one flavonol: quercetin.[9] This was an interesting outcome because previous studies had shown negligible absorption and rapid clearance of pure quercetin in humans.[11] Surprisingly, Hertog's group discounted any other phenolic compounds, even other flavonoids, as having any potential effect on CHD.[9]

Candidate Compounds for Nutritional Activity

Wine is a particularly rich dietary source of flavonoid phenolics, so many studies to uncover a cause for wine's effects have focused on its phenolic constituents, particularly resveratrol and the flavonoids.

The phenolic compounds are well known to oenologists for their sensory properties, and for this reason their chemistry has been investigated in wine for decades. These substances give wine its bitterness and astringency, and are the foundation of long ageing, since they are effective antioxidants. Wines low in these substances, such as white wines, rarely age gracefully.

Resveratrol. Resveratrol has been studied intensely of late, partly because it was only in 1992 that E.H. Siemann and Leroy Creasy at Cornell University first described it in wine.[12] The researchers focused attention on resveratrol by suggesting that it could reduce CHD mortality, a contention based on the compound's ability to inhibit platelet aggregation and reduce lipid levels in hyperlipidaemic rats. These effects were noted in studies to elucidate the activities of piceid, a glucoside of resveratrol, and the main component of "Koji jon," a traditional Chinese medicine prepared from the roots of *Polygonum cuspidatum* and used to treat atherosclerosis.[13] More recently, resveratrol has also been shown to inhibit the oxidation of human low density lipoproteins (LDL).[14] In addition, another study has shown that resveratrol, as well as some flavonoids, inhibits eicosanoid synthesis, a key step in platelet aggregation.[15]

This flurry of research activity has been augmented by the more recent discovery of resveratrol's *cis* isomer and piceid in grapes and wine.[16] To date at least ten different chromatographic methods have been described to analyse wine and/or grapes for resveratrol. Perhaps the latest is a liquid chromatographic method that separates the *cis* and *trans* isomers of both resveratrol and piceid.[16]

The low levels of resveratrol in wine, averaging 1-2mg/l *trans*-resveratrol and perhaps 5mg/l for all derivatives, has not abated the interest in it. An estimate of the maximum resveratrol blood levels that could arise from wine consumption—making a number of generous assumptions, including the consumption of a substantial amount of wine (500ml) that contains high levels of resveratrol (5mg/l), and efficient absorption (50%) and slow metabolism, for a 75kg individual with 5 litres blood—is about 1μM. This is enough to attain the lowest resveratrol concentration known to elicit the *in vitro* effects mentioned above, typically 1-10μM. Certainly, the levels needed *in vivo* may be different from those seen to be effective *in vitro*, but based on these assumptions, the likelihood that resveratrol from wine would have a physiological effect on a wine drinking population appears to be low.

Flavonoids. Of course there are many other phenolic phytochemicals in wine that occur at a much higher concentration than resveratrol. Some of the most abundant are found at more than 100mg/l. These are mainly the flavonoids, which include four major classes: the flavonols, the anthocyanins, the catechins or flavonols, and oligomers (procyanidins) and polymers (tannins) of the catechins. The total amount of these compounds present is 1-3g/l for red wines and 0.2g/l for whites. In addition, wine contains a significant amount of non-flavonoids, including the hydroxy-cinnamates, benzoic acids, stilbenes and others, typically 0.2-0.4g/l for all wines. The number of individual phenolic compounds in a particular wine varies, but a typical liquid chromatographic analysis of red wine will reveal 50 constituents—closer scrutiny will reveal many more.

In many situations, but not all, these phenolic compounds are antioxidants. To oversimplify the disease of atherosclerosis, or clogging of the arteries, the oxidation of LDL is an important step in the development of arterial plaque,[17] and so substances that can block this oxidation should slow the disease. In fact, some antioxidants have slowed atherosclerosis in animals. Thus the question arises, could wine phenolics affect atherosclerosis in humans? This issue has been addressed in a recent issue of *Biochemist* dedicated to the question of the nutritional value of antioxidants.[18]

A key study of wine by Edwin Frankel and colleagues at the University of California, Davis,[19] showed that wine contained antioxidants towards the oxidation of LDL *in vitro*. Thus it is possible that these substances inhibit LDL oxidation *in vivo* and, as a result, slow the development of arterial plaque. Based on this interpretation, these authors suggested that the phenolic compounds in wine were responsible for the French paradox. A follow-up investigation on three wine phenolics showed that all were antioxidants for LDL. The most potent were epicatechin and quercetin, while resveratrol was less potent, and the control, α-tocopherol (vitamin E) was least so.[14]

These *in vitro* studies are interesting but do not establish whether or not the oxidation of LDL is inhibited by wine antioxidant constituents *in vivo*. Two studies have attempted to answer this question indirectly. One showed that wine consumption increased blood antioxidant activity.[20] Unfortunately, this method analysed serum which was prepared by first clotting the blood sample. Since wine phenolics inhibit clotting,[15,21] it is likely that the measured change in antioxidant capacity was at least in part a result of differential clotting. The other study showed that daily wine consumption reduced the tendency of isolated LDL to oxidise.[22] Unfortunately, in both these investigations, the chemical composition of the wine used was not characterised, and so it is impossible to ascribe the observed effects to any specific components.

Since the aggregation of platelets (thrombosis) is an important factor in precipitating a heart attack, compounds that reduce platelet activity could also reduce CHD mortality. Wine flavonoids have been show to inhibit platelet aggregation.[21] They appear to do this by specifically inhibiting oxygenase enzymes.[23] It also seems that there are significant differences in the ability of the different flavonoids to affect platelets—quercetin is potent, while catechin is not.[15]

The significance of these *in vitro* studies has recently been complemented by *in vivo* studies using a well developed animal thrombosis model. Cyclic flow reductions are used in animals as a model for thrombosis—chemicals that reduce the cyclic flow reductions are thought to lower the chance of platelet aggregation and hence thrombosis. The consumption of wine or flavonoids greatly attenuated cyclic blood flow reductions.[24] Since grape juice also decreased the blood flow reductions (albeit with three times as much volume) it appears that non-alcoholic constituents are the active components. It is unknown whether the different potencies are the result of altered levels of phenolics or an effect of ethanol.

The Absorption of Phenolics

Phenolics must be absorbed into the blood stream to have any direct effect on coronary diseases, but data in the field of flavonoid absorption by humans are very sparse. In the available data, there is a striking difference between two of the compounds that have been well studied, catechin and quercetin.[25] About half the administered dose of catechin was absorbed, while there was no detectable absorption of quercetin. Clearly there are profound differences in the behaviour of these two compounds, and at present there is no explanation for this. In addition, if quercetin is not absorbed from wine or other foods by humans, then its nutritional significance would be very limited.

Dietary Sources

Wine is certainly not the only dietary source of phenolic compounds. Fresh fruits are a rich source and, notably, Criqui and Ringel found that fruit consumption correlated highly with reduced CHD mortality.[4] Wine phenolics all originate from the grape, although wine making does alter the constitution. Tea is another potent phenolic source, but green and black tea have quite different compositions. Green tea mainly contains the monomeric catechins while black tea has the oligomeric and polymeric forms. To date, studies of tea's effects on CHD mortality have not distinguished between green and black tea,[26] but east Asian countries where green tea consumption is high are noted for their low CHD mortality rates.[3]

While some of the phenolic compounds are common to many plants, there are many differences as well. These differences have been exploited in the authentication of juices and wines. For instance, *Vitis vinifera*, the wine grape, contains almost exclusively the monoglucosides of the anthocyanins while American grape species and crosses contain the diglucosides.

Compared with other dietary sources, wine contains relatively high levels of phenolics. It is also a diverse source, as it includes significant amounts of all the major classes of phenolics. While whole fruits are rich sources, the data on juices show low levels. Apparently, normal aerobic processing degrades these compounds.[27] Wine production, on the other hand, is largely anaerobic, and thus the phenolic compounds are retained. The ethanol produced during fermentation is an effective solvent to extract the substantial amount of flavonoid phenolics in both the skins and the seeds present in red grapes. In white wine production the skins and seeds are separated from the juice immediately after crushing the grapes, and thus flavonoid levels are much lower. However, the phenolics present solely in the white juice, mostly the non-flavonoids, are retained in anaerobic wine production.

Conclusions

The combination of epidemiological, clinical and mechanistic studies strongly suggests that wine phenolics are beneficial nutrients that can reduce CHD mortality. In order to make sound dietary recommendations for phenolic compounds and the foods that contain them, such as wine, much work remains to be done. The availability, absorption, metabolism and physiological effects of well-described chemical systems need to be thoroughly investigated. The striking differences between two flavonoids in absorption and effects on eicosanoid synthesis are a reminder that these compounds do not represent a simple class of substances with similar behaviours. A detailed understanding based on sound chemistry will not only clear up a current paradox, but may clarify the effects of other phenolic-rich foods, notably the anti-cancer activity of tea. If future studies bear out these suspicions, chemists will be able to toast "to your health" with enthusiasm and understanding!

References

1. Goldberg, D.M., *Clin. Chem.*, 1995, **41**, 14-6
2. St. Leger, A.S., Cochrane, A.L., & Moore, F., *Lancet*, 1979, **i**, 1017-20
3. Renaud, S., & de Lorgeril, M., *ibid*, 1992, **339**, 1523-6
4. Criqui, M.H., & Ringel, B.L., *ibid*, 1994, **344**, 1719-23
5. Marmot, M.G., Rose, G., & Shipley, M.J., *ibid*, 1981, **i**, 580
6. Rimm, E., *et al, ibid*, 1991, **338**, 464-8
7. Handa, K., *et al, Am. J. Card.*, 1990, **65**, 287-9
8. Klatsky, A.L., & Armstrong, M.A., *ibid*, 1993, **71**, 467-9
9. Hertog, M.G.L., *et al, Lancet*, 1993, **342**, 1007-11
10. Hertog, M.G., *et al, J. Agric. Food Chem.*, 1992, **40**, 2379-83
11. Gugler, R., *et al, Eur. J. Clin. Pharmacol.*, 1975, **9**, 229-34
12. Siemann, E.H., and Creasy, L.L., *Am. J. Enol. Vitic.*, 1992, **43**, 49-52
13. Arichi, H., *et al, Chem. Pharm. Bull.*, 1982, **30**, 1766-70
14. Frankel, E.N., *et al, Lancet*, 1993, **341**, 1103-4
15. Pace-Asciak, C.R., *et al, Clin. Chimica Acta*, 1995, in press
16. Lamuela-Raventós, R.M., *et al, J. Agric. Food Chem.*, 1995, **42**, 281-3
17. Esterbauer, H., *et al, Free Radical Commun.*, 1992, **13**, 341-90
18. Leake, D., *Biochemist*, 1995, **17**, No. 1, 12-15
19. Frankel, E.N., *et al, Lancet*, 1993, **341**, 454-7
20. Maxwell, S., Cruickshank, A., & Thorpe, G., *ibid*, 1994, **344**, 193-4
21. Seigneur, M., *et al, J. Appl. Card.*, 1990, **5**, 215-22
22. Fuhrman, B., *et al, Am. J. Clin. Nutr.*, 1995, **61**, 549-54
23. Mower, R., *et al, Biochem. Pharmacol.*, 1984, **36**, 317-22
24. Demrow, H.S., Slane, P.R., & Folts, J.D., *Circulation*, 1995, in press
25. Hackett, A.M., in "Plant flavonoids in biology and medicine: biochemical pharmacological and structure-activity relationships", (Eds. V. Cody, E.J. Middleton & J.B. Harborne), *New York: Liss*, 1986, 177-94
26. Klatsky, A.L., *et al, Ann. Epidemiol.*, 1993, **3**, 375-81
27. Spanos, G.A., & Wrolstad, R.E., *J. Agric. Food Chem.*, 1990, **38**, 1565-71 ∎

Questions

1. What is the "French Paradox," and what dietary factor was used to explain it?

2. What effect was moderate alcohol consumption found to have on blood lipid levels?

3. How might antioxidants slow the progression of atherosclerosis?

Answers begin on page 161.

21 *Dimethyl ether, CH$_3$-O-CH$_3$, is the simplest of the ethers. It can be readily prepared from natural gas, CH$_4$, or from synthesis gas, which is a mixture of carbon monoxide and hydrogen. Dimethyl ether (DME) has a high volatility and is especially flammable. Because of its ultraclean burning characteristics, DME could become an alternative to diesel fuel. The use of DME shows great promise in eliminating two of the major problems of conventional diesel fuels: emission of nitrogen oxides and soot. It would be a welcome change for the environment to eliminate the black smoke that sometimes pours from diesel-burning trucks and cars.*

Amoco, Haldor Topsoe Develop Dimethyl Ether as Alternative Diesel Fuel

A. Maureen Rouhi

Dimethyl ether use in diesel engines cuts soot and nitrogen oxides emissions, reduces engine noise.

Dimethyl ether (DME) is shaping up as an ultraclean alternative fuel for diesel engines, according to two companies working to develop synthetic fuels from natural gas. Use of DME in diesel engines results in lower emissions of nitrogen oxides (NO$_x$), practically soot-free combustion, and reduced engine noise. And emission levels from diesel engines run on DME will meet and even surpass California's ultra-low-emissions vehicle (ULEV) regulations for medium-duty vehicles such as commercial trucks and buses.

These findings were presented at this year's international congress and exposition of the Society of Automotive Engineers International, held in Detroit. They are the outcome of joint efforts by Amoco, a major producer of natural gas, and Haldor Topsoe, a Copenhagen-based company engaged in chemical process development, engineering, and catalyst manufacture.

The simplest of ethers, DME currently is used mainly as an environmentally friendly propellant for spray cans. Unlike chlorofluorocarbons, it is not harmful to the ozone layer. The ether is nontoxic and easily degrades in the troposphere.

Because of DME's narrow use, current world production capacity is only about 150,000 metric tons per year, according to Amoco and Haldor Topsoe. But DME may become a big-time commodity. Since the late 1980s, Haldor Topsoe has been trying to expand DME's uses, says John B. Hansen, manager of the company's catalytic processes department. Spurred by the oil price hikes of the late 1970s and early '80s, the company in 1982 embarked on a development program for production of synthetic gasoline from synthesis gas via methanol and DME. The result was TIGAS: Topsoe integrated gasoline synthesis.

TIGAS "was a big technological success," says Hansen, because of the large degree of integration of process steps. "But when we finalized the project, oil prices were down to $10 per barrel, from $30 per bbl when we started. So [TIGAS] was not interesting anymore from the economic point of view.

"We thought about how to use what we had learned during that project and came up with the idea of using DME—which is formed in TIGAS—as an intermediate in other syntheses. And then one of my lab technicians [Svend-Erik Mikkelsen] got this crazy idea of testing this fuel in his lawn mower and it worked. Then we went on to a forklift in the company. That worked also, even without spark plugs connected. We decided to get professional help for the motor side of it and started this collaboration," initially with the Technical University of Denmark, Copenhagen, and then with Amoco.

Preliminary studies were carried out by Spencer C. Sorenson, professor of mechanical engineering at the Technical University of Denmark, and Haldor Topsoe's Mikkelsen. Their work shows DME has the potential to eliminate the major problems of conventional diesel fuel: soot and NO_x. They tested neat (99.9+%pure) DME in a small (0.273L), direct-injection, nonturbocharged diesel engine. For these studies, they did not try to optimize the fuel system, but they pressurized it because of DME's volatility.

They found that DME is comparable with diesel fuel in terms of thermal efficiency—the efficiency of converting chemical energy to mechanical energy. DME and diesel fuel also were comparable with respect to hydrocarbon and carbon monoxide emissions. But engine noise was lower with DME. And in terms of soot and NO_x emissions, DME gave significantly better results. NO_x emissions from DME were only about 25% of those from diesel fuel. And "for all intents and purposes, the black soot is gone," says Sorenson.

A major problem with diesel engines is emission of NO_x and particulate matter, primarily soot. NO_x emissions are particularly undesirable. They contribute to acid rain, formation of ground-level ozone in urban areas, and depletion of stratospheric ozone.

With conventional diesel fuel, there is a trade-off between NO_x emissions and soot, Sorenson explains. Soot can be reduced but at the expense of higher NO_x emissions. "That [trade-off] is eliminated with DME," he says. Because combustion is soot free, "we now have a free hand to work with the NO_x emissions."

Sorenson and Mikkelsen also tested a simulated "raw" DME—92% DME, 4% water, and 4% methanol—which would be cheaper to make. In terms of thermal efficiency, soot, and NO_x emissions, the lower grade fuel was comparable with neat DME. But hydrocarbon and CO emissions varied depending on load. With the engine working hard (high load), the two fuels were comparable. But at idling conditions (low load), impure DME gave more CO and hydrocarbons than did the neat fuel.

To further reduce NO_x emissions, the researchers tried exhaust gas recirculation (EGR). With 30% EGR, NO_x emissions were reduced 90%, with no loss in thermal efficiency and no soot but with a 70% increase in CO emissions.

With these promising initial results, Haldor Topsoe asked Amoco to look at DME. Since the late 1970s, Amoco has been engaged in development work for synthetic fuels from natural gas. "DME was one of our candidates, but it was too costly to produce," says Theo H. Fleisch, Amoco's manager for gas transportation and upgrading. Haldor Topsoe's studies showing that DME can be produced at relatively low cost in large plants rekindled Amoco's interest in DME.

Amoco directed rigorous testing of DME with a bigger engine. The Amoco team, led by Fleisch, worked with researchers from Haldor Topsoe, Navistar International Transportation in Chicago, and AVL Powertrain Engineering in Novi, Mich. The tests were carried out at facilities of AVL List in Graz, Austria.

The team used a 7.3-L turbocharged direct-injection diesel engine, the type used extensively in trucks and buses. To ensure that DME remains liquid at all operating and environmental conditions, a simple fuel storage and delivery system was designed. The team also modified the fuel injection system to allow delivery of higher volumes of DME and installed an air-air intercooler to maintain temperatures at acceptable levels. But the engine was otherwise not modified.

The results far exceeded expectations. At all engine speeds and loads tested, combustion with DME was soot free. This allowed use of high EGR flows, resulting in very low NO_x emissions. Even without exhaust treatment, emissions from DME were below the 1998 California ULEV regulations for medium-duty vehicles. DME emissions also satisfied most proposed European requirements for trucks. But improvements were needed to meet 1996–2000 standards for NO_x and CO emissions for U.S. passenger cars.

The group also tested DME diluted with up to 10% water and 10% methanol in equal parts. With increasing dilution, soot and NO_x emissions decreased, whereas hydrocarbon and CO emissions increased. But overall, the data indicated that relatively large amounts of dilution can be tolerated while still meeting ULEV regulations. Further emissions reductions can be gained, the group proposed, by fine-tuning engine parameters, in particular, by optimizing the fuel injection system.

Engineers Paul Kapus and Herwig Ofner of AVL List took on that task. Their job was to establish parameters for the fuel injection equipment and the combustion system. Because soot is not an issue with DME, they focused on minimizing NO_x emissions while optimizing thermal efficiency and keeping combustion quiet. They showed that liquid injection of DME is possible at pressures below 250 bar. Therefore, fuel injection systems for dedicated DME engines can be less expensive than those for conventional diesel engines, says James C. McCandless, president and CEO of AVL Powertrain Engineering. Conventional diesel engines normally operate at pressures of 1,200 to 2,000 bar.

Thus, not only does DME burn better and cleaner than conventional diesel fuel, but the likely cost of redesigned engines to run on it also will be reasonable. "[DME] is probably one of the most exciting things that has happened in alternative fuels. For once we seem to have an engine and a fuel that are really well-suited for each other," says Norman D. Brinkman, principal, research engineer at General Motors' fuels and lubricants department.

And one more factor in DME's favor: It can qualify as a renewable fuel, because synthesis gas, a mixture of carbon monoxide and hydrogen, can be produced from biomass. But the cheapest raw material for synthesis gas will be natural gas.

DME currently is produced in two steps: formation of methanol from synthesis gas and dehydration of methanol to DME. With Haldor Topsoe's TIGAS, the two steps are combined into one process, called oxygenate synthesis. In effect, three reactions are taking place simultaneously in one reactor:

$$2H_2 + CO \rightarrow CH_3OH$$
$$CO + H_2O \rightarrow CO_2 + H_2$$
$$2CH_3OH \rightarrow CH_3OCH_3 + H_2O$$

The first two reactions make up a typical methanol synthesis. But the equilibrium does not favor methanol formation. Thus, on its own, methanol synthesis must be carried out at high pressures (80 to 120 bar). Adding the third reaction relieves the thermodynamic constraints by turning methanol into DME. The strong synergistic effects allow high conversion of synthesis gas in a single pass, according to Hansen.

To make the process efficient, Haldor Topsoe had to develop dual-function catalysts. A mixture of classical methanol and DME catalysts wouldn't have been useful, because at the high temperatures expected for oxygenate synthesis such catalysts usually yield excessive by-products, mainly higher alcohols and hydrocarbons.

Haldor Topsoe has been demonstrating the technology with a pilot plant in Copenhagen. The plant produces 50 kg of raw DME per day. The company reports the catalyst is stable—the original catalyst charge has been operating for more than 12,000 hours. Catalyst selectivity also is good, with only very small amounts of higher alcohols and aliphatic hydrocarbons being formed.

If indeed DME becomes accepted as a substitute for conventional diesel fuel, demand for it will rise, and new production facilities will have to be developed. Haldor Topsoe's analyses show that the most economical route to DME is large stand-alone plants based on natural gas. The company claims it can build facilities with capacities up to 7,000 metric tons of DME per day.

Amoco and Haldor Topsoe estimate that DME will be more expensive than conventional diesel fuel. But Amoco's Fleisch believes consumers can be convinced that the fuel deserves a premium price. "Retail customers will not pay for ultralow emissions," he says. "They will pay, however, for higher fuel efficiency, lower maintenance, and overall better performance."

DME seems to be a winner. Indeed, the Big Three automakers, the Department of Energy, and two California government agencies have expressed "a significant interest" in the technology, according to AVL Powertrain's McCandless.

But there are problems. "You have to have plants that can make large quantities of the fuel," says Brinkman. "That's difficult but probably not as difficult as having an infrastructure—the pipelines, the trucks that haul the fuel, and the retail outlets that distribute the fuel. This is a significant barrier for all alternative fuels."

And the public has to be convinced to use DME. "You can't force a fuel on society," says Fleisch. "A market pull needs to develop from customers, original equipment manufacturers, and government agencies." That pull will come if customers accept DME. Thus, he says, "We want the public to know DME is customer friendly, combines the high fuel efficiency of diesel engines with ultralow tailpipe emissions, and can be cost effective." ■

Questions

1. What problems are associated with conventional diesel fuel?

2. Why is dimethyl ether (DME) potentially better than diesel fuel?

3. What effects do nitrogen oxide emissions have on the environment?

Answers begin on page 161.

22 *Formaldehyde, the simplest aldehyde, is a gas at room temperature. It is a key industrial chemical in the production of plastics such as Bakelite and Formica. Formaldehyde is also used in the preparation of adhesives and fillers, in the manufacture of plywood and fiberboard, and in the production of insulating materials. Unfortunately, formaldehyde is also an irritant and a potential air pollutant in homes.*

Clearing the Air

There's some evidence, as we reported in February, that houseplants, especially the sturdy and ubiquitous spider plant, can reduce formaldehyde in indoor air slightly. This is good news, but you shouldn't expect potted plants to do all the work. You may remember formaldehyde as a smelly liquid you encountered in high-school biology lab, with a frog or some other preserved specimen in it. One of its hundreds of uses is, of course, as embalming fluid for everything from lab specimens to humans. It is a natural substance—a normal by-product of living cells. Low levels of it occur in foods, though such tiny amounts have no effect in the body. But, as emitted from manufactured goods in its gaseous form, it is a pollutant of indoor air.

One of the top 50 industrial chemicals in this country, formaldehyde is utilized in thousands of products, notably in the resins or glues used in particleboard, fiberboard, and hardwood plywood (the glues in softwood plywood are also made with formaldehyde but emission levels are lower). It is often an ingredient in plastic products and some cosmetics and drugs. It is present in car exhaust, cigarette smoke, and emissions from gas stoves and furnaces, kerosene heaters, and power plants, as well as in permanent-press fabrics and urea-formaldehyde foam insulation materials, which were widely used in Canadian and U.S. homes after 1973. It's impossible to avoid formaldehyde entirely.

The worse-case scenario for exposure would be to live in a new mobile home where the flooring, cabinets, and furnishings made of hardwood plywood and other pressed-wood products, to have a closet full of brand-new permanent-press clothing and permanent-press draperies at every window and new carpets on the floor, and to keep the doors and windows closed. Mobile homes tend to be worse than other homes because they not only contain a lot of new pressed-wood materials, but enclose them in a very compact space. New products are worse than older ones, since emissions lessen with time. High humidity and temperatures can increase emissions.

What Can It Do to You?
A normal level of formaldehyde in indoor and outdoor air is considered less than 0.03 parts per million and causes no problems. But when it rises above 0.1 parts per million, it may produce skin irritation, watery eyes, burning sensations in the eyes and respiratory tract, coughing, and similar reactions. It affects some people more than others. It is hard to tell what's causing these symptoms, which may also be caused by other pollutants or a cold. Workers exposed to high levels of formaldehyde have been shown to be at increased risk for some cancers, especially of the nose and throat. High levels also cause cancer in laboratory animals.

The Occupational Safety and Health Administration (OSHA) has sought tighter restrictions on formaldehyde exposure in the workplace. The Environmental Protection Agency and Consumer Product Safety Commission are also implementing changes to reduce human exposure. The Department of Housing and Development (HUD) requires building materials to meet certain standards for formaldehyde emissions. Generally, the use of formaldehyde in such consumer products as permanent-press fabrics, pressed woods, and floor coverings has been reduced, and products containing more than 1% formaldehyde have to be labeled.

What Can You Do about It?
If you have symptoms that may be related to formaldehyde exposure, consider having your home checked. The California Air Resources Board in Sacramento lists two monitors suitable for measuring formaldehyde: one (about $35) from Air Technology Labs in Fresno, California (phone 800-354-

Reprinted by permission from the *University of California at Berkeley Wellness Letter,* © Health Letter Associates, 1994.

2702); the other (about $75 for two monitors) made by Air Quality Research International in Durham, North Carolina (919-544-2987).

If you don't have symptoms, you needn't go to this expense. But the steps below, which minimize formaldehyde exposure, are worth taking even if you have no symptoms:

• **Increase home ventilation,** especially in warm weather. In the cold months, don't seal up all the windows and doors. Circulating fresh air is necessary to control indoor pollution and allergens of all kinds.

• **Wash permanent-press clothing and sheets** before you use them. (These fabrics do contain less formaldehyde than they used to.) If you're installing permanent-press draperies that you can't wash, try to air them out for a few days before hanging them.

• When buying hardwood plywood and other pressed woods for home building projects, **get materials stamped with the HUD emissions seal.**

• **If you buy unfinished furniture or other pressed-wood products, varnish or paint them.** A waterproof finish such as polyurethane will greatly reduce formaldehyde emissions.

• **If you already have formaldehyde foam insulation, seal it off.** Patch any cracks, or cover the walls with nonporous wallpaper. Seal wood paneling with varnish.

• **If you buy a mobile home, be sure an adequate ventilation system is installed.** HUD regulations require a seller to give a buyer an information sheet on ventilation before a sales agreement is reached.

• **Provide good ventilation in a motor home or other recreational vehicle,** especially in warm weather, and be sure to air out the vehicle after it has been in storage. RVs, which don't fall under HUD jurisdiction, may contain hardwood-plywood or particleboard fittings, as well as new curtains and upholstery, all potential sources of formaldehyde. ∎

Questions

1. Why do mobile homes tend to have higher concentrations of formaldehyde in the air than other types of housing?

2. What can be done to reduce emissions from pressed-wood products and unfinished furniture?

3. What suggestions are given concerning recreational vehicles?

Answers begin on page 161.

23 *Years ago, if you had a headache or fever, you went to the store for a bottle of aspirin. Today, the leading commercial pain reliever is still common aspirin, but there are dozens of over-the-counter preparations from which to choose. Despite the variety of packaging and advertising promises, all nonprescription pain relievers contain one of four substances: aspirin, acetaminophen, ibuprofen, or naproxen sodium. Knowing the advantages and disadvantages of each type of pain reliever is important.*

Pain, Pain Go Away

Ruth Papazian

Used to be, aspirin and other salicylates were the only medications available for nonprescription relief of minor ailments—from headaches and fever to muscle strain and minor arthritis. Today, consumers looking for temporary relief from such garden-variety ills have their pick of what can be a bewildering array of "regular," "extra-strength," and "maximum pain relief" tablets, caplets and gel caps on the drugstore shelf.

Though this cornucopia can seem confusing, the products' pain-relieving ingredients fall into just four categories: aspirin (and other salicylates), acetaminophen, ibuprofen, and naproxen sodium. For the most part, these over-the-counter (OTC) analgesic ingredients are equally effective. How-ever, some may be more effective for certain types of ailments, and some people may prefer one type to another because of their varying side effects. "Knowing the pros and cons of each type of pain reliever will allow you to choose among them," says William T. Beaver, M.D., professor of pharmacology and anesthesia at George-town University School of Medicine in Washington, D.C.

Old Faithful

Americans have been reaching for aspirin for almost 100 years as an all-purpose pain reliever (see "Aspirin: A New Look at an Old Drug" in the January-February 1994 *FDA Consumer*). Aspirin (or acetylsalicylic acid) works in part by suppressing the production of prostaglandins, hormone-like substances that have wide-ranging roles throughout the body, such as stimulating uterine contractions, regulating body temperature and blood vessel constriction, and helping blood clotting. "Regular" strength aspirin contains 325 milligrams (mg) per tablet; "extra" or "maximum" strength, 500-mg per tablet. The usual adult (defined as 12 years and older) dosage is one to two 325-mg aspirin tablets every four hours.

Some manufacturers add caffeine to aspirin. "There is no evidence that caffeine relieves pain, but it can enhance the effects of aspirin, possibly by lifting a person's mood," says Michael Weintraub, M.D., director of FDA's Office of OTC Drug Evaluation. Since a two-tablet dose provides roughly the same amount of caffeine as a cup of coffee, you can get the same effect by taking two plain aspirin with coffee.

To minimize the stomach irritation aspirin can cause, some brands are "buffered" with calcium carbonate, magnesium oxide, and other antacids or coated so the pills don't dissolve until they reach the small intestine. Buffered formulas may offset aspirin's directly irritating effects on the stomach lining. They may be useful for people who get heartburn or stomach pain when they take aspirin, as well as for those with arthritis, who need to take as much as 4,000 mg every day.

Aspirin also causes gastrointestinal (GI) upset indirectly (by inhibiting production of a prostaglandin that protects the stomach lining by stimulating mucus production); buffering does nothing to offset this effect.

The downside of coated aspirin products is that they may take up to twice as long to provide pain relief as plain aspirin, according to Weintraub. Last September, an FDA advisory panel recommended that labels on products containing aspirin warn that heavy drinkers are especially vulnerable to developing GI bleeding.

Aspirin should not be taken by people who have:
• ulcers, because it can worsen symptoms

Reprinted from *FDA Consumer*, January–February 1995, pp. 11–14.

- asthma, because it can trigger an attack in some asthmatics
- uncontrolled high blood pressure, because of an increased risk of one type of stroke
- liver or kidney disease, because it may worsen these conditions
- bleeding disorders or who are taking anticoagulant medication, because it may cause bleeding.

Continual high dosages of aspirin can cause hearing loss or tinnitis—a persistent ringing in the ears.

FDA requires products containing aspirin and other salicylates to carry a label warning that children and teenagers should not use the medicine for chickenpox or flu symptoms because of its association with Reye syndrome, a rare disorder that may cause seizures, brain damage, or death.

The label also alerts pregnant women that use of aspirin in the last trimester may increase the risk of stillbirth and of maternal and fetal bleeding during delivery.

One Aspirin Alternative

Twenty years ago, FDA approved acetaminophen (Tylenol, and other brands and generics) in dosages of 325 mg and 500 mg for OTC use. "Nobody knows exactly how acetaminophen works, but one theory is that it acts on nerve endings to suppress pain," says Weintraub. Acetaminophen is as effective as aspirin in relieving mild-to-moderate pain and in reducing fever, but less so when it comes to soft tissue injuries, such as muscle strains and sprains, he adds. The usual adult dosage is two 325-mg tablets every four hours.

Acetaminophen-based products to ease menstrual cramps often contain other ingredients, such as pamabrom (a diuretic) or pyrilamine maleate (an antihistamine used for its sedative effects). "While these ingredients are safe, they have not been proven effective against uterine cramps, although they may relieve other symptoms associated with menstrual pain," says Weintraub.

Though acetaminophen is no better or faster at pain relief than aspirin, the drug is gentler on the stomach and reduces fever without the risk of Reye syndrome. However, even at moderate doses, acetaminophen can cause liver damage in heavy drinkers. At press time, FDA was planning to require a warning about this on the labels of OTC products containing the drug.

From Rx to OTC

Like aspirin, ibuprofen and naproxen sodium inhibit prosta-glandin production. However, they are more potent pain relievers, especially for menstrual cramps, toothaches, minor arthritis, and injuries accompanied by inflammation, such as tendinitis. FDA approved ibuprofen for OTC marketing in 1984 at a dosage level of 200 mg every 4 to 6 hours, and naproxen sodium in 1994 at a dosage level of 200 mg every 8 to 12 hours.

"Ibuprofen and naproxen sodium were converted to OTC status after their manufacturers did the necessary studies to show that these pain relievers were effective at OTC dosages, which are lower than prescription dosages," explains Weintraub. The lowest dosage strength for prescription-strength ibuprofen (Motrin and others) is 300 mg per tablet, and 275 mg per tablet for the prescription version of naproxen sodium (Anaprox, for example). "In addition, the pharmaceutical companies had to show that these drugs were safe for use by a larger, more varied group of people [than would have received them by prescription only] and that the drugs were safe to use without medical supervision, as is the case with all nonprescription drugs."

Taken at the recommended adult dosage, OTC ibuprofen (Advil and others) and naproxen sodium (Aleve) are somewhat gentler on the stomach than aspirin. However, people who have ulcers or who get GI upset when taking aspirin should avoid both. In addition, asthmatics and people who are allergic to aspirin should avoid ibuprofen and naproxen sodium. An FDA advisory panel has recommended labeling on ibuprofen products like that recommended for aspirin, warning heavy drinkers about increased risk of gastric bleeding and impaired liver function (products with naproxen sodium labels already include this information).

Although ibuprofen and naproxen sodium interfere with blood clotting much less than aspirin does, they should not be used by people who have bleeding disorders or who are taking anticoagulants. Children under 12 should not be given either drug, except under a doctor's supervision, and people over 65 are advised to take no more than one naproxen sodium tablet every 12 hours.

Choosing an OTC pain reliever involves balancing effectiveness for a particular ailment with side effects. Often this is a very individual choice, based in part on your health history and how the drug affects you. Regardless of which type of OTC pain reliever you choose, remember that it is intended to be used on a short-term basis, unless directed by a doctor, cautions Weintraub. The warning labels on these products include limitations on duration of use to ensure that chronic or serious illnesses are not masked. Typically, labels advise against taking the product for more than 10 days to relieve pain (for children, the upper limit is five days), or more than three days to reduce fever. If symptoms worsen, pain persists, or there is redness or swelling, medical attention should be sought. ■

Questions

1. What is a drawback of coated aspirin products?

2. What advantages does acetaminophen have over aspirin?

3. What is the difference between the prescription and over-the-counter versions of ibuprofen and naproxen sodium?

Answers begin on page 161.

24 *Americans spend upwards of $20 billion each year on unproven medical treatments. Approximately 80% of older Americans have one or more chronic health problems. Their pain and disability lead to despair, making them excellent targets for deception. "Soothing . . . strong . . . trustworthy"—slick advertisements hold out hope for the elderly that the next quick "cure" will work. Concerned about the health of a parent or child, family members, too, encourage their loved ones to "try everything," especially unproven remedies. Compounding the problem are drugstores that are like supermarkets, selling anything that is profitable. Mixed in on the shelves with genuine remedies are a growing number of questionable and sometimes risky products.*

Drugstore Deceptions: Separating Hype from Hope

Many self-care remedies offer only empty promises. They're a waste of money—and some are risky as well.

People depend on their pharmacist as a healthcare professional—an expert in the risks and benefits of medications. The products stocked on pharmacy shelves tend to take on some of the pharmacist's own credibility. Increasingly, however, the drugstore has turned into a small department store, stocking whatever is profitable. Among the more profitable items mixed in with the genuine remedies are a growing number of questionable, sometimes risky, and often overpriced products. Here's a rundown of some notable offenders.

Supplements: A to Zinc
Enter just about any drugstore and behold a dizzying array of nutritional supplements. One widely distributed brand, for example, boasts some 200 different vitamin formulas, including 18 versions of vitamin C. To round out their product lines, many manufacturers offer "special formulas" of dubious value at best. Among the most dubious:

"Natural" vitamins. Manufacturers would have you believe that "natural" is somehow better. But it's just more expensive. Despite the enchanting image conjured by "rose hips," the body can't tell a natural molecule of vitamin C from a synthetic one.

Stress formulas. Extreme physical stress—like running a marathon—can increase the body's need for certain nutrients, but even then rarely enough to warrant adding supplements to a balanced diet. Emotional stress has no effect whatever on your body's vitamin needs, nor do vitamins have any effect on your ability to withstand such stress. Legal actions by the Federal Trade Commission and several states' Attorneys General have stopped most vitamin manufacturers from making explicit anti-stress claims for their products. But implied claims still abound. Lederle Laboratories, for example, says a lot with just one word: *Stresstabs*. The familiar slogan—"for people who burn the candle at both ends"—drives the point home. The product features hefty amounts of the water-soluble B vitamins. They're likely neither to help nor to harm; but they'll waste your money in return for nothing more than bright-yellow urine.

Iron-fortified formulas. If you saw the recent movie "Quiz Show," you may have been reminded of those bygone years of cloying *Geritol* commercials that touted the high-iron supplements as a cure for "tired blood." Yes, there is a fatigue-causing condition called iron-deficiency anemia. But it's typically found only in women who menstruate especially heavily or among people bleeding internally from an undetected source. Attempts to cure fatigue by taking iron can mask a serious underlying condition. Moreover, getting too much iron may increase the risk of cancer and coronary heart disease (see CRH, 7/94).

Bottle of the sexes. Special vitamins for men and women (*One-A-Day for Men, One-A-Day for Women, Women's Changing Times*, and so on) are gimmicks. The differences between men's and women's nutritional needs are generally insignificant. Some women may need greater amounts of calcium than men do. But since calcium is a bulky mineral, you won't find a

---

"Drugstore Deceptions: Separating Hype from Hope." Copyright 1995 by Consumers Union of U.S., Inc., Yonkers, NY 10703-1057. Reprinted by permission from *Consumer Reports on Health*, March 1995.

multiple supplement—gender-specific or unisex—that contains the full U.S. RDA of 1000 milligrams anyway. (Postmenopausal women who are not on estrogen and all men over 65 should get at least 1500 mg calcium a day, whether from diet or pills.)

Weight-Loss Losers
In 1993, weight-conscious Americans spent over $50-million on diet pills at the drugstore, but few got their money's worth. The most common active ingredient is phenylpropanolamine (PPA), which supposedly suppresses appetite by stimulating the central nervous system. (The drug is also the active ingredient in the decongestant *Propagest*.) In 1982, based on extremely slim data, a Food and Drug Administration panel classified PPA as safe and effective for weight loss. But the agency has been back-pedaling ever since, and insiders say it's only because of bureaucratic chaos that PPA-based diet pills have yet to be yanked from the shelves.

A recent survey of CONSUMER REPORTS readers found that of those who had tried PPA-based diet pills like *Acutrim* and *Dexatrim*, fewer that 5 percent were satisfied. That's probably because, as one leading weight-loss researcher told us, the drug typically results in "trivial" weight loss, if any. And the risk of serious side effects—including abnormal heart rhythms and raised blood pressure—is substantial.

The other big seller among diet pills is the fiber-based variety, which is also a dud. Manufacturers of fiber pills have come under repeated fire from the Federal Trade Commission for making misleading claims. But like ducks in a shooting gallery, as soon as one "miracle" fiber product gets knocked off the shelves, another pops up.

Take the case of Schering's *Fibre Trim*. Last summer, under pressure from the FTC, the manufacturer agreed (without admitting guilt) to stop making claims about the weight-loss benefits of the product until it could provide adequate scientific proof. Meanwhile, Schering had taken the product off the market altogether. Since then, another manufacturer, California-based KCD, Inc., has introduced another fiber product: *SeQuester*. According to the package, S*eQuester* "reduces fat from the food you eat . . . effects [sic] metabolizable energy . . . [and] negatively affects the availability of fat and sugar." If they're trying to say that the stuff helps you lose weight, they're wrong.

It is true that fiber can help lower cholesterol levels (see Article 28, "Fiber Bounces Back"), but cholesterol has nothing to do with flabby thighs. It's also true that fiber can expand in the stomach and may thereby reduce hunger pangs. But you'd have to consume 23 *SeQuester* tablets (at a cost of about $3.50) to match the fiber content of one apple.

The latest contenders to weigh in are the so-called reducing creams. In two tiny, unpublished studies, creams containing the asthma drug aminophylline did appear to reduce fatty thighs. But the effect was slight, and the risks are unknown. Some researchers fear that the creams could harm the circulatory system. Nevertheless, the allure of a magic slimming cream was enough to inspire manufacturers to rush a slew of products to market—including *Cellution, Skinny Dip, Smooth Contours*, and others—all targeted at thigh-conscious women. Not to neglect male concerns, Nobel Pharmaceuticals introduced a similar product for men. It's called *Belly Buster*.

Incredible Shrinking Hemorrhoids
Hemorrhoid ointments, if they help at all, usually do so by providing lubrication to let stools slide over the distended veins with a minimum of friction and irritation. As for the fancy ingredients—aloe, shark oil, and the like—that pad the bill, there's no evidence that they'll help you sit in peace. Although the FDA allows products containing blood-vessel constrictors—including market-leader *Preparation H*—to claim that they'll "shrink" hemorrhoids, our medical consultants scoff at the claim. No remedy can do that.

The FDA has banned some of the more questionable ingredients in certain hemorrhoid remedies—including a substance called live yeast-cell derivative, which had been the key ingredient in *Preparation H* and some other products. In response, the manufacturers changed their formulations. Our experts view those fixes as frivolities that the FDA hasn't taken the time to reject. If the agency does crack down again, the formulation will likely change again.

The latest act to join the hemorrhoid circus is a product called *Hemorid*, manufactured by Florida's Thompson Medical Co., Inc., a giant in the diet-pill field. *Hemorid*, as model and television spokeswoman Kim Alexis explains, is a hemorrhoid medicine "made just for women." Sure enough, it comes in a lovely pink box. For that, you pay extra: A 1-ounce tube of *Hemorid* costs about $6—roughly 30 percent more than the already pricey *Preparation H.* Color scheme aside, there's nothing about *Hemorid* that's especially suited to women.

Most hemorrhoid suffers can stop their suffering through a variety of self-help measures—such as following a high-fiber diet, cleaning the anal area carefully, taking warm baths, and using a simple water-based ointment like *K-Y Jelly*. In rare cases when self-help strategies fail to relieve hemorrhoid pain, the topical anesthetic pramoxine may provide some additional—though temporary—relief. *Hemorid* has it, but you'll find cheaper, single-ingredient ointments, such as *Tronolane* (about $4 for the same 1-ounce tube). (For more on self-care and medical treatments for hemorrhoids, see CRH, 2/94).

Rembrandt: A New Master?
Rembrandt Whitening toothpaste is part of a product line that calls itself "the new standard in oral care." The toothpaste certainly sets a new standard in price. At about $8 for a 3-ounce tube, it's over eight times more expensive than familiar products like *Colgate* or *Crest.* Is *Rembrandt* toothpaste really worth the premium?

Dentist Robert Ibsen, president of Den-Mat Corp., the California company that makes *Rembrandt*, tells us his product "represents a significant advance in tooth-cleaning technology." According to a recent *Rembrandt* ad, the special low-abrasion formula "safely whitens teeth while reducing plaque, tartar, and extrinsic stain—even better than *Crest.*"

To support the claim about plaque and tartar, the company sent us two small, short studies published a few years ago in two non-research dental journals. Our dental consultants were not impressed and have not seen any confirming evidence in a peer-reviewed research journal. There's also a paucity of published data on *Rembrandt's* safety. Some experts express concern that, over time, the paste's unusual enzyme-containing formula might irritate soft tissue and damage exposed tooth roots.

Rembrandt does indeed appear to be less abrasive than most other toothpastes. But abrasiveness isn't usually a concern with toothpaste. "I don't know of any major product with a problem level of abrasion," says Clifford Whall, Ph.D., of the American Dental Association. Only people who have had extensive cosmetic dental work might want to shell out extra on a super-low-abrasion toothpaste.

And *Rembrandt's* whitening ability? In 1992, CONSUMER REPORTS tested the cleaning power of some four dozen toothpastes. *Rembrandt* scored 53 on a 100-point scale, indicating moderate success at removing stains. But 33 of the tested brands—including *Crest*—scored higher. The report concluded that "*Rembrandt* lagged far behind many conventional and far cheaper brands."

Homeopathy in Hiding
Last March, CONSUMER REPORTS concluded that there's "no logical scientific case to be made for homeopathy's benefits." Whatever you think of homeopathy, it stands to reason that homeopathic products should be clearly labeled as such. But clear labeling is evidently a low priority for the manufacturers of *Vagisil Yeast Control* and *Yeast Gard* vaginal suppositories, two products that promise to relieve the burning and itching of vaginal yeast infections. You'll find them stocked alongside two other vaginal yeast fighters, *Gyne-Lotrimin* and *Monistat 7.* The boxes all seem confusingly similar on the outside, but the stuff inside couldn't be more different.

Gyne-Lotrimin and *Monistat 7* are the over-the-counter versions of two drugs—clotrimazole and miconazole—that were until recently available only by prescription. Both have been clinically proved to cure yeast infections. *Vagisil Yeast Control* and *Yeast Gard*, on the other hand, are "alternative" medicines that contain immeasurably small amounts of fungi and an herb. Their efficacy is entirely unproved. Most medical experts doubt even the theoretical basis for homeopathy—the notion that substances that would otherwise provoke symptoms can, when diluted to the vanishing point, actually cure disease.

Most homeopathic remedies are at least openly marketed as such, but these two "anti-yeast" products state only in small type that they are homeopathic preparations; we think they'll snare consumers who might otherwise buy bona-fide medications. The products represent what could be the start of a disturbing trend. The FDA has generally not required homeopathic remedies—unlike other types of drugs—to be proved safe and effective. It should. Otherwise, unscrupulous vendors will be able to mislead unwitting consumers who don't read the fine print. ■

Questions

1. How does emotional stress affect the body's need for vitamins?

2. Why can too much iron intake be dangerous?

3. What serious health risks are associated with phenylpropanolamine (PPA), the active ingredient in many weight-loss pills?

Answers begin on page 161.

25 *Caffeine is an alkaloid that is present in coffee, tea, cola drinks, chocolate, and cocoa. It is a mild stimulant of the respiratory and central nervous systems, the reason for its well-known side effects of nervousness and insomnia. During the last 20 years, there have been many conflicting reports about caffeine's influence on heart disease, cancer, and miscarriages. Can people continue to enjoy their favorite blends in moderation without undue risk?*

Caffeine: Grounds for Concern?

Caffeine has been one of the most studied substances, but the results of the research have often been confusing. At one time or another, caffeine has been accused of causing miscarriages, pancreatic cancer, heart disease, breast disease, high blood pressure, and high blood cholesterol levels. But does it cause any of these? First some other questions—and answers.

What Is Caffeine? Why Do People Like It?

Caffeine, one of the most popular and most ancient stimulants, occurs in more than 60 cultivated plants and trees. It is one of a group of compounds called methylxanthines, which stimulate certain neurotransmitters in the central nervous system.

People like caffeine because, for many of them, it wards off drowsiness, increases alertness, and shortens reaction time. It helps millions wake up and feel better in the morning. It helps them stay awake on the road and gives them something to drink socially that isn't alcohol. It may even increase aspirin's effectiveness as a painkiller, which is why it is added to some pain relievers. Some studies have found that caffeine improves reading speed, enhances performance on math and verbal tests, and produces an increased capacity for sustained intellectual effort in general. This is especially true in subjects who are tired or bored, in children as well as adults. Other studies, however, have found that caffeine does not significantly affect verbal fluency, numerical reasoning, or short-term memory.

Does It Have Physical Benefits?

Over the years, coaches and athletes have reported beneficial effects of caffeine for endurance exercise such as cycling or cross-country skiing. Caffeine boosts athletic performance supposedly by helping the body break down fats for use as energy, thereby sparing glycogen for later use and delaying exhaustion. So far, however, research has produced inconsistent results. Several small studies have shown that caffeine in two or three cups of brewed coffee (220 to 330 milligrams), taken within a few hours of endurance exercise, postponed exhaustion in well-conditioned athletes. But other studies have found little or no benefit from caffeine. Like so many effects of caffeine, the effect on exercise varies from person to person. If you're a regular coffee drinker, don't expect coffee to have dramatic effects on your athletic performance; to get a boost, you would have at abstain from caffeine for three or four days and then consume some before the event. If you're not accustomed to caffeine, be careful, since you may have an adverse reaction while exercising.

What Are the Adverse Effects?

Depending on how much you consume, caffeine can temporarily step up your heart rate and increase stomach-acid secretion and urine production. Such effects are minimal among healthy adults who consume moderate amounts of caffeine—about two or three cups of coffee a day. The most common of caffeine's ill effects is "coffee nerves"—trembling, nervousness, insomnia, muscle tension, irritability, headaches, and disorientation. This usually occurs only in people who don't normally consume caffeine. Your reaction depends not only on your habituation and sensitivity to caffeine, but also on the amount you consume, your body weight and physical condition, and your anxiety level in general. Children run a special risk of coffee nerves, since one cola for a small child may have the same effect as two to four cups of coffee for an adult. But children who regularly consume caffeine are less affected than children who do not.

Many people who drink caffeinated coffee regularly have no trouble sleeping after a late-night

Reprinted by permission from the *University of California at Berkeley Wellness Letter*, © Health Letter Associates, 1994.

cup. But those unaccustomed to it may have trouble falling asleep, followed by disturbed sleep patterns. Caffeine levels peak in your body within an hour of consumption; more than half the caffeine is metabolized (that is, broken down and rendered inactive) in three to seven hours. Thus an afternoon cup of coffee or tea is unlikely to keep anyone awake eight hours later. Of course, psychology plays a role, too: if you expect that something will keep you awake, you may not sleep well.

What about Caffeine and Heart Disease?

Scores of studies have failed to establish any conclusive link to coronary artery disease (CAD). Many studies that claimed to find such a link were flawed, often because confounding factors were ignored. Recently, a review of studies on the subject in *Heart & Lung* found no increased risk of heart attack in people who consume caffeine, even if they had heart disease. In fact, the researchers suggested that cardiac patients *not* have their caffeine restricted while in the hospital (up to five cups of caffeinated beverages a day). In addition, a recent report in the *Archives of Internal Medicine* found no association between coffee drinking—even more than six cups a day—and CAD.

Caffeine may boost blood pressure in those not accustomed to it, but only temporarily. Still, people with hypertension (diastolic blood pressure above 105) should ask their doctors about steering clear of caffeine. There's little or no evidence that caffeine increases irregular heartbeats, or arrhythmias. But people subject to them, particularly if recovering from a heart attack, should also get medical advice.

But Doesn't Coffee Boost Cholesterol?

The effect of caffeine on blood cholesterol is murky. For every study that finds that caffeine boosts cholesterol, another finds the evidence inconclusive and several discover no effect at all. A study at the Johns Hopkins Institutions in 1992, which found that healthy men who drank four cups of filtered coffee daily experienced a slight rise in total blood cholesterol, received lots of press. But HDL ("good") cholesterol rose proportionately with LDL ("bad") cholesterol, which means that the men's risk of heart attack was not increased. Two Dutch studies did show that four to six daily cups of unfiltered European-style coffee raised blood cholesterol levels, while filtered coffee did not. Why this should be is a mystery, but the researchers theorized that a standard paper filter (which most Americans use) may trap some cholesterol-raising substance, whatever it may be.

Why the Discrepancies among the Studies?

Caffeine's effects are hard to study because the response to caffeine varies considerably from person to person. In addition, many of the discrepancies can be explained by the fact that researchers have not consistently distinguished between habitual coffee consumers and non-coffee-drinking volunteers who ingest large doses of caffeine over the course of an experiment. A dose of 250 milligrams given to a regular coffee drinker usually has no significant effect on blood pressure, heart rate, respiration, metabolic rate, blood cholesterol, or anxiety level. But caffeine given to subjects who haven't been consuming caffeine for a week or so can temporarily raise all these. Another complication in evaluating caffeine's impact on health is that people who drink a lot of coffee also tend to smoke, exercise little, eat lots of fat, and have other habits that put them at risk for a variety of ailments. Keep up good health habits, and coffee drinking is seldom a problem.

What about Benign Breast Disease? Cancer?

The fibrocystic tissue that forms in the breasts of some women can be lumpy and painful. Some doctors advise women with fibrocystic breasts to limit their intake of caffeine. But several major studies have found no relationship between caffeine consumption and fibrocystic breasts. Nor is there any evidence that giving up caffeine by itself eases the discomfort some women experience.

What about cancer? Despite reports of links between coffee drinking and cancer of the bladder, pancreas, and breast, studies have failed to confirm them. The authors of a 1981 study suggesting a link between pancreatic cancer and coffee reversed their findings in a second study fives years later. Last June, a study in the *Lancet* found no link between bladder cancer and coffee drinking. Also in 1993, the Iowa Women's Health Study found no link between caffeine and breast cancer.

Does Caffeine Cause Stomach Problems?

Don't blame caffeine. Coffee does stimulate the flow of stomach acid and thus potentially irritates the stomach lining and ulcers. It may also relax the sphincter at the end of the esophagus (as do alcohol and smoking), allowing for the backup of stomach contents and thus increasing the chance of heartburn in some people. However, decaffeinated coffee seems to be almost as much of a problem as regular coffee. The principal culprits are either the beans' natural oils or substances apparently introduced during the roasting process. So people with ulcers or chronic heartburn should avoid both regular coffee and decaf. Even caffeine-free herbal teas and grain-based beverages can stimulate stomach acid.

Does Caffeine Interfere with Nutrient Absorption?

Not caffeine, but other substances (such as polyphenols) in coffee and especially tea may interfere with the body's absorption of certain minerals, notably calcium and iron—especially nonheme iron, the kind found in plant foods. This is only rarely a problem. If you drink tea with every meal or drink many cups per day while eating a strict vegetarian diet, it's possible that the tea could promote iron deficiency, but only if you're eating minimal amounts of iron-containing foods. Tea and coffee in moderation can be part of a healthy diet, and there's no evidence that they lead to iron-deficiency anemia or calcium loss.

Should You Kick the Habit?

If caffeine gives you a lift, there's no reason to deprive yourself of its benefits—unless you're pregnant or trying to conceive. Clearly, if you think caffeine may be robbing you of a sound night's sleep, by all means try cutting it out in the evening. And if you get jittery and nervous from it at any time of day, it makes sense to cut back.

About going cold turkey: Since caffeine is a mildly habit-forming drug, most coffee and tea drinkers experience headaches and other withdrawal symptoms, such as irritability and nausea, when they quit. Such symptoms, which usually start 12 to 16 hours after their final dose and last for up to six days, can be avoided simply by cutting back on caffeine gradually. ∎

The Daily Dose

This chart will help you calculate your daily caffeine intake. But remember, caffeine content varies widely, depending on the product you use and how it's prepared.

	Caffeine (mg)
Coffee, drip or brewed, 6 oz	80–175
Coffee, instant, 6 oz	60–100
Coffee, decaffeinated, 6 oz	2–5
Tea, 5-minute steep, 6 oz	20–100
Hot cocoa, 6 oz	2–20
Coca-Cola, 12 oz	30–45
Milk chocolate, 1 oz	1–10
Bittersweet chocolate, 1 oz	5–35
Chocolate cake, 1 slice	20–30
Anacin, Empirin, or Midol, 2 pills	64
Excedrin, 2 pills	130
NoDoz, 2 pills	200

Questions

1. How will drinking 1.5 to 3 cups of coffee per day affect the health of pregnant women?

2. Why are caffeine's effects difficult to study?

3. Other than caffeine, what substances in coffee and especially in tea interfere with nutrient absorption?

Answers begin on page 161.

26 *With today's growing environmental concerns, many chemical manufacturers are rethinking existing preparations for a variety of chemical products. For example, several important commercial products can be prepared from glucose rather than benzene, the usual starting material. Unlike benzene, glucose is nontoxic and is obtained from renewable resources like plant starch and cellulose. The effort to incorporate environmental consciousness into chemical research and development is known as* green chemistry *or* green technology.

Green Chemistry at Work

John Frost

Products can be made from glucose instead of benzene.

Although many people may never have heard of benzene, everyone has come in contact with the materials, flavors, or medicinal agents that come from manufacturing processes using benzene as a starting material. Vanillin, a dominant flavor component of vanilla ice cream, is derived from benzene, as is hydroquinone, a chemical essential to image formation in photography. Nylon 66, a synthetic fiber widely used in fabrics, is made from benzene. Benzene is the starting material for synthesis of drugs such as L-DOPA, used to treat Parkinson's disease. Phenol, catechol, and pyrogallol are examples of "building block" molecules derived from benzene that are employed in chemical manufacture.

The 1.7 billion pounds of benzene produced each year in the United States provide one measure of its utility. At the same time, there are a number of environmental reasons for avoiding the use of benzene in chemical manufacture. Perhaps most compelling: Benzene is a potent carcinogen.

Scrutiny of many of the aforementioned chemicals derived from benzene reveals that each molecule contains at least one oxygen atom while benzene completely lacks oxygen atoms. Introduction of oxygen to make up for this lack can require processes that are environmentally problematic. One of the steps used to introduce oxygen atoms during manufacture of adipic acid, a component of Nylon 66, is responsible for 10 percent of the annual global increase in atmospheric nitrous oxide. This byproduct is a causative agent of atmospheric ozone depletion and has been implicated in global warming. Conversion of benzene into catechol, a chemical essential to vanillin manufacture, requires use of concentrated hydrogen peroxide. This oxidant is a highly energetic, corrosive material requiring special care during handling and storage.

Also, benzene is primarily obtained from petroleum. All of the environmental costs, such as from oil spills, associated with use of this nonrenewable resource must be factored into the true environmental cost of using benzene as a starting material. With support from EPA and the National Science Foundation, alternative manufacturing processes are being explored. By these new methods, chemicals usually created from benzene are made instead from nontoxic glucose, a component of table sugar.

Nontoxic glucose has six oxygen atoms attached to six carbon atoms. Use of this highly oxygenated starting material eliminates the step during manufacture of adipic acid that generates nitrous oxide. The route developed for conversion of glucose into catechol eliminates use of corrosive hydrogen peroxide. Processes using glucose as the starting material typically employ water and temperatures no higher than body temperature. This procedure contrasts with the elevated temperatures and organic solvents often used in chemical manufacture where benzene is the starting material. Waste streams for use of glucose as starting material are typically no different than what is normally handled by municipal sewage treatment facilities. This is not necessarily the case with traditional chemical manufacture. Unlike benzene, glucose is obtained from such renewable resources as plant starch and cellulose.

The key to employing glucose in chemical manufacture is the use of biocatalysis. Enzymes found in

Reprinted from *EPA Journal*, Fall 1994, pp. 22–23.

laundry detergents are one type of biocatalyst; the intact microbes used to make beer represent another. The processes developed to convert glucose into chemicals such as adipic acid, catechol, and hydroquinone employ intact microbes. Normally, glucose is consumed by microbes and then "burned," producing carbon dioxide and providing the energy needed for growth and reproduction. This process is very similar to the production of energy (i.e., heat) when wood is burned.

By altering a microbe's metabolism, glucose that would normally be burned can instead be channeled into biosynthetic pathways used by the microbe to make chemicals. A biosynthetic pathway consists of a series of enzymes that are located inside the microbe. By isolating DNA fragments from one type of microbe and then introducing these DNA fragments into another microbe, biosynthetic pathways can be created that do not normally exist in nature. This is the process used to create the microbial biocatalysts that can convert a solution of sugar (glucose) water into a solution of adipic acid, catechol, or hydroquinone.

"Green" manufacturing routes ideally should lead to chemicals that are economically competitive with chemicals produced by traditional methods. For two chemicals of roughly comparable cost, the consumer or producer can then be realistically expected to choose in favor of the chemical produced by a "green" process. Projections indicate that catechol and hydroquinone can be biocatalytically produced from glucose at a cost competitive with current market prices. Although the estimated manufacturing cost for adipic acid exceeds the market price, the costs of eliminating emissions of benzene and nitrous oxide will put upward pressure on the cost of current manufactures of adipic acid from benzene.

Deriving chemicals from glucose presents numerous scale-up and processing problems. The best crops for manufacture of chemicals may not be those that currently dominate American agriculture–such as corn. More appealing are plants like switchgrass, that can be harvested multiple times during a growing season and that require minimal fertilizing and pesticide inputs. Additional challenges confront the grain processing companies that control the renewable resources and glucose supply. These commercial entities typically have little experience in traditional chemical markets. At the same time, chemical companies lack a significant presence in renewable resources and typically lack experience in biocatalysis.

These barriers to change in the manufacture of chemicals are imposing, although powerful incentives for surmounting such barriers are generated by the need for compliance with increasingly strict governmental regulations and the need to fulfill public environmental expectations. Synthesis of chemicals from glucose using biocatalysis offers the promise of achieving fundamental environmental improvement while increasing the demand for agricultural products. In the final analysis, what is good for the environment can also be good for American agriculture. ∎

Questions

1. Name at least two substances that can be produced using benzene.

2. What health risk does benzene pose?

3. What is the key to employing glucose as a substitute for benzene in chemical manufacturing?

Answers begin on page 161.

27 *Sugar (sucrose) is the most commonly used food additive in the United States. Each U.S. citizen consumes an average of 100 pounds of sugar per year. Two-thirds of this amount results from products to which the manufacturers have added sugar. Sugar has not been viewed favorably in the press. Numerous accusations have been made against sugar, but this article may help put sugar in a more balanced perspective.*

What's Wrong with Sugar?

Sugar's reputation as a health heavy is much overblown.

Over the years, sugar has been blamed for ills ranging from obesity to hyperactivity—even psychotic rage. Fear of sugar has created an industry devoted to sugar substitutes and has turned *Cokes* into guilt trips. A look at the facts may help put sugar into a more realistic health perspective.

Weight Gain: Thin Evidence
Sugar is not nutritious. It contributes only "empty" calories to the diet. And those calories can add up. If you drank three 12-ounce cans (160 calories each) of sugar-sweetened soda a day and otherwise maintained your usual diet and exercise habits, you could gain about a pound a week. The desire to avoid extra calories has fed the demand for artificial sweeteners (see box).

But sugar is not the only caloric culprit or even the main one in most sweets. Fat shares the blame. A chocolate eclair, for example, gets 133 calories from fat and only 55 from sugar. Even a chocolate candy bar gets more calories from fat (122) than from sugar (92). And the calories in a cup of vanilla ice cream are just about evenly divided between the two (roughly 130 calories apiece).

Indeed, many overweight people suffer less from a "sweet tooth" than from a "fat tooth." Observational studies have found that overweight people actually eat *less* sugar than thin people—but more dietary fat.

Hyperactivity: No Defense
For years, parents and teachers have observed that children "bounce off the walls" after eating sweets. That observation has led to the belief that sugar causes hyperactivity. In adults, the same logic has been taken to its extreme with the now infamous "*Twinkie* defense"—the contention that certain murderers were driven to their violent behavior by sugary junk food. Numerous studies have discredited those notions.

Earlier this year, researchers published in The New England Journal of Medicine the strongest case yet against the alleged link to hyperactivity—findings that were described as "resoundingly negative" by an accompanying editorial. The study involved some 50 children, ages 3 to 10, half of them believed by their parents to be sensitive to sugar. For one week the children and their families ate specially prepared meals high in sugar; during two separate weeks, the same meals were prepared with artificial sweeteners. The participants weren't told which was which—and they couldn't tell the difference.

At the end of each week, the researchers gathered data on 39 measures of the children's behavior, mood, and thinking ability. They found no sign of hyperactivity after the high-sugar meals. Even the parents who believed their children usually reacted to sugar didn't notice any effect. So if sugary treats make kids bouncy, it's due to the treat, not the sugar.

If anything, sugar may have the opposite effect. Some of the children in that study showed slight improvements in learning tasks and slightly slower times in tests of speed and agility when on the high-sugar diet. The authors interpreted those findings as evidence of a modest *relaxing* influence from sugar, a possibility that has more scientific basis than theories about hyperactivity. As we noted in our May report on foods and performance, at least a half dozen clinical trials suggest that a meal high in carbohydrates such as sugar makes people feel somewhat more relaxed or even sluggish.

Blood Sugar: Not So Low
Some people believe that sugar triggers symptoms of hypoglycemia, or low blood sugar, such as sweating, trembling, rapid heartbeat, and

"What's Wrong with Sugar?" Copyright 1994 by Consumers Union of U.S., Inc., Yonkers, NY 10703-1057. Reprinted by permission from *Consumer Reports on Health,* October 1994.

Sugar Is Sweet and So Are Substitutes
The appeal of artificial sweeteners lies largely in their lack of calories, a benefit that leads dieters to put up with the artificial taste. But research suggests that many would-be dieters sabotage that potential benefit by loading up on other high-calorie treats as a reward for using "diet" products.

Still, artificial sweeteners can help control calories so long as the rest of the diet is under control. And unlike sugar, the artificial sweeteners don't promote cavities or affect blood-sugar levels in diabetics. Here's a rundown of the three artificial sweeteners now on the market and three more on the horizon.

Saccharin (*Sugar Twin, Sweet 'N Low*), the oldest artificial sweetener, is 300 times sweeter than table sugar. It's used in baked goods, candy, canned fruit, chewing gum, dessert toppings, jams, and soft drinks. (It's also widely used in medications and vitamins.) In 1977, the U.S. Food and Drug Administration proposed banning saccharin because animal studies suggested a cancer risk. Under pressure from soft-drink makers and calorie-conscious consumers, Congress placed a moratorium on the proposed ban and mandated a warning label instead. In 1991, the FDA quietly withdrew its proposal. The warning label remains.

Aspartame (*Equal, NutraSweet, NatraTaste*), 180 times sweeter than sucrose, is widely used in candy, cereals, frozen desserts, and soft drinks. Although some commercial baked goods contain aspartame, it shouldn't be used for most home cooking because it's unstable at high temperatures. As one of the most thoroughly tested additives ever approved by the FDA, aspartame appears to be entirely safe—almost: People with the inherited disorder phenylketonuria (PKU), which affects roughly 1 in 15,000 Americans, must limit their intake of the amino acid phenylalanine, a component of aspartame. For that reason, aspartame-containing products bear a warning aimed at people with PKU.

Acesulfame K (*Sunette, Sweet One*), 200 times sweeter than sucrose, is used in dry beverage mixes, candy, chewing gum, gelatins, and puddings. It's the only sugar substitute that bears no warning label of a possible health hazard.

The following sweeteners are currently seeking FDA approval:

Cyclamate, 30 times sweeter than table sugar, was widely used as an artificial sweetener in the 1950s and '60s, but was banned in the U.S. as a potential carcinogen in 1970. The manufacturer and a trade association are now urging the FDA to rescind the ban.

Sucralose, 600 times sweeter than table sugar, has shown no health risks in more than 100 animal and human studies conducted over the past 15 years.

Alitame is 2000 times sweeter than table sugar. The FDA has requested more studies to bolster the dozen or so supporting the sweetener's safety.

---

hunger. But those symptoms are rarely due to sugar.

True, eating lots of sugar can sometimes cause blood-sugar levels to drop a bit. That's because the pancreas responds to a hefty dose of sugar by secreting the hormone insulin, which helps the body's cells draw sugar out of the bloodstream. In a few people, that can push blood-sugar levels below normal. But such drops in blood sugar almost never cause symptoms. The rare exceptions occur in people with a condition called reactive hypoglycemia. For the vast majority of people who believe their symptoms are caused by sugar consumption, there's actually some other cause, such as an overactive thyroid or panic attacks. People who suspect hypoglycemia should have their physician help determine if their symptoms coincide with a drop in blood sugar.

A different sort of hypoglycemia is more common among diabetics. But in those people, hypoglycemia has nothing to do with eating too much sugar. Instead, it happens only when they inject too much insulin (or take too much oral antidiabetic medication), eat too little food, or exercise too much.

Diabetes: Off the Hook
Traditionally, physicians have warned diabetics to avoid or strictly limit sugary foods—not because of resulting low blood-sugar levels, but to control the abnormally *high* blood-sugar levels that characterize the disease and damage the eyes, kidneys, nerves, and circulatory system. Now even that prohibition has been repealed.

Earlier this year, the American Diabetes Association issued revolutionary dietary guidelines based on the contention that sweets are no more upsetting to a diabetic's blood-sugar levels than other carbohydrates such as bread, pasta, and potatoes, The new guidelines call for a balance between carbohydrates and fat, the two main dietary components.

One legacy of the longtime prohibition against sugar in diabetics is the popular belief that sugar somehow *causes* diabetes. It doesn't. The real culprits are problems with the body's production of insulin and its sensitivity to the hormone.

Decay: One Charge with Teeth

The only real risk from sugar for most people is tooth decay. While fluoride, plastic sealants for children's teeth, and improved oral hygiene have dramatically reduced the incidence of cavities, it's still important to watch what you eat—and when you eat it (see CRH, 4/93).

All carbohydrates—starches as well as sweets—can be converted into tooth-dissolving acid by the bacteria in dental plaque. Since the simple sugars in sweets are most readily available to the bacteria, they pose a special risk. But starches can be equally harmful if they stick around long enough for their complex carbohydrates to get broken down into simple ones. The most damaging foods seem to be sweet baked goods like cookies and cakes, which cling to the teeth and contain both quick-acting sugars and slower starches. Sugar-sweetened products that dissolve slowly, such as hard candy and throat lozenges, can also prolong the acid attack.

To minimize the cavity risk from sugars, wait for mealtime to consume carbohydrates. That way, other foods will increase saliva production, which neutralizes the acids and helps clear food particles and sugar from the mouth. When you want a snack, choose fruit or raw vegetables instead of a sweet or starchy snack. Although the sugar in fruit can cause cavities, most raw fruits are safe to snack on because they're relatively low in sugar and aren't very sticky. The exceptions are bananas, which are quite high in sugar, and dried fruits, such as dates and raisins, which are both sugary and sticky.

Natural Appeal: Artificial Argument

Many people believe that sucrose—table sugar—is somehow less healthy than so-called natural sugars, like honey or fruit sugar. Many food companies cash in on that mistaken belief with an array of "naturally sweetened" products ranging from cereals and yogurts to cookies and candies. In some cases, that just boosts the overall sugar content of the food. For example, a bowlful of *Honey Nut Cheerios* contains 11 grams of sugar to plain *Cheerios'* 1 gram.

In any case, sugar is sugar. "Natural" sugars are no more nutritious than sucrose. Like sucrose, natural sweeteners such as honey, brown sugar, and corn syrup contain only calories and nothing else worth measuring. (The one exception is blackstrap molasses, which contains calcium, iron and some other nutrients.) While fruit juices do contain many nutrients, the tiny amount used in fruit-juice-sweetened products has no significant nutritional value.

Natural sugars provide just as many calories as sucrose and are at least as bad for the teeth. In fact, syrupy sweeteners like honey may be worse because they're sticky. And contrary to what many diabetics have been led to believe, products sweetened with fruit juice have no less an impact on blood-sugar levels than foods sweetened with sucrose. ∎

Questions

1. What results did the *New England Journal of Medicine* report about a study in which sugar was given to children?

2. Explain how eating lots of sugar causes blood-sugar levels to drop.

3. Why shouldn't aspartame be used for most home cooking?

Answers begin on page 161.

28 *Dietary fiber is obtained from various sources, including the outer covering (bran) of wheat and other grains, the pectin of apples and other fruits, and the skins and fibrous parts of vegetables and fruits. Ingested fiber passes through the small intestine essentially unchanged. The most important function of dietary fiber is to provide bulk, which helps the body get rid of waste. For several years, there has been a great deal of public interest in the fiber content of foods. Research continues to show that a high-fiber diet might lower cholesterol levels and reduce the risk of heart attack and colon cancer.*

Fiber Bounces Back

After all the jokes about oat bran, fiber is getting the last laugh: It does help fight deadly disease.

In the late 1980s, reports that oat bran might protect the heart by lowering cholesterol levels spawned a craze for oat-bran muffins and transformed the image of dietary fiber from mere "roughage" to life-saving substance. When a highly publicized study from Harvard failed to confirm that benefit, fiber in general and oat bran in particular became a national laughing-stock—a symbol of spurious health fads and flip-flopping research that encourages people to ignore studies and just eat whatever they want.

Now the pendulum has swung back in fiber's favor. And this time the shift is not based merely on the results of the latest study but rather on a substantial body of research showing that fiber can indeed lower cholesterol levels, protect the heart, and probably reduce the risk of cancer.

Fiber and Coronary Heart Disease
Observational studies of large populations have repeatedly shown that people who eat a high-fiber diet are less likely to develop coronary disease than those who get little dietary fiber. Several of those studies were designed to minimize the chance that the reduced risk might be due to features of a high-fiber diet other than the fiber itself. For example, researchers from England and California evaluated nearly 900 middle-aged and older people for 12 years. After adjusting for the consumption of fat, calories, alcohol, calcium, protein, and several other nutrients thought to affect the heart, the researchers found that coronary deaths dropped by 25 percent for every 6-gram increase in daily dietary fiber—roughly the amount in a half-cup serving of baked beans.

That apparent link between fiber and reduced risk has been bolstered by studies of one of the two basic types of fiber, the type that dissolves in hot water. Soluble fiber, plentiful in beans, oat bran, most fruits, and some vegetables, inactivates certain digestive acids made from cholesterol in the liver; in response, the liver draws cholesterol from the blood to make more acid. In addition, gastrointestinal bacteria ferment soluble fiber, creating chemicals that may slow the liver's production of cholesterol. In theory, those effects may help reduce cholesterol levels and, in turn, the risk of coronary disease.

The highly publicized Harvard study of oat bran that seemed to refute any such fiber-cholesterol connection turned out to have serious flaws in design. Researchers subsequently reviewed the results of seven clinical trials in which all dietary changes were carefully tracked, to isolate any cholesterol-lowering effect of soluble fiber itself. The combined results of those studies, which included some 850 people, showed that consuming just 3 extra grams of soluble fiber per day—roughly the amount in half a serving of oat-bran cereal or one serving of beans—can lower cholesterol levels by an average of 6 mg/dl.

That's just a 3 percent drop for someone with an average cholesterol level. But the reduction in those trials occurred almost entirely in the harmful LDL (low-density-lipoprotein) cholesterol, not in the beneficial HDL (high-density-lipoprotein) cholesterol. An increase of 5 grams in soluble-fiber intake per day—the amount recommended for someone eating the average American diet—would typically cut cholesterol levels by 5 to 10 percent; that translates into a 10 to 20 percent reduction in coronary risk. People with above-average choles-

"Fiber Bounces Back." Copyright 1995 by Consumers Union of U.S., Inc., Yonkers, NY 10703-1057. Reprinted with permission from *Consumer Reports on Health*, March 1995.

terol levels could expect even larger reductions in both cholesterol and risk.

Fiber and Coronary Risk Factors

Some evidence suggests that fiber, particularly soluble fiber, may reduce not only cholesterol levels but also the risk or severity of three other major risk factors for coronary disease:

Diabetes. Populations that consume a lot of fiber generally have a lower risk of diabetes than those that get little fiber. Further, clinical trials have consistently shown that fiber helps lower blood-sugar levels in people who have already developed the disease. That benefit is probably due to soluble fiber, which slows the absorption of sugar from the small intestine.

Obesity. Soluble fiber mixes with liquids in the stomach to form a gelatinous mass, making a meal feel larger and linger there longer. In theory, that might help overweight people control their weight by curbing their appetite. Several clinical trials comparing similar weight-loss regimens have shown that adding fiber to the diet helps people lose an average of four additional pounds over a two-to-three-month period. People who get the extra fiber by switching from a typical American diet to one rich in fruits, vegetables, grains, and beans should shed more weight than that, since those foods are not only high in fiber but also low in calories and fat.

Hypertension. Two large observational studies from Harvard University—one involving some 60,000 women and the other, some 30,000 men—have tried to isolate the effects of fiber on hypertension by controlling for various nutrients thought to influence blood pressure. The men who ate the most fiber were 46 percent less likely to develop high blood pressure than those who ate the least; women who ate the most fiber had 24 percent less risk of hypertension. However, it's not yet clear whether fiber actually accounts for that apparent protection, since clinical trials have failed to confirm the benefit.

Fiber and Cancer

The strongest evidence that fiber may hep prevent malignancies comes from research on colon cancer. According to one review, 29 of 37 observational studies have linked a high-fiber diet with a reduced incidence of colon cancer. Attempts to confirm that link have focused on insoluble fiber, abundant in whole grains, beans, most vegetables, and some fruits. Insoluble fiber doesn't become gelatinous, but it does eventually absorb water and swell, adding bulk to the contents of the colon. In theory that may reduce the risk of cancer by speeding potentially cancer-causing wastes through the colon and by increasing the size of the stools, thereby reducing the concentration of those wastes.

Feeding insoluble fiber to animals lowers their risk of developing colon cancer. In humans, several small clinical trials have shown that consuming insoluble fiber reduces the number of precancerous cells lining the colon and rectum. Another small trial found that such fiber inhibits formation of polyps, mushroom-shaped growths in the colon that can turn cancerous. Two much larger trials testing fiber's effect on the development of polyps are currently underway.

Limited evidence suggests that fiber may also lower the risk of breast cancer. In animals, insoluble fiber reduces levels of the female hormone estrogen, which is suspected of stimulating breast-cancer growth. The first major observational study on fiber and breast cancer in humans did not find any connection between fiber and reduced breast-cancer risk. But the next one did—and that study, involving some 60,000 Canadian women, controlled for several potentially confounding factors that were not addressed in the earlier study. A few smaller investigations have also connected fiber with reduced risk of the disease.

The Fiber Message

Overall, studies have clearly strengthened the notion that fiber not only fights relatively minor intestinal disorders, such as diverticulosis and constipation, but also helps prevent coronary disease and probably cancer. However, the strongest evidence of a major reduction in risk comes from studies of diets that are rich in high-fiber *foods,* not just in fiber alone. Such foods, low in fat and rich in nutrients, have many health benefits beyond those provided by fiber. So regardless of whether studies ever nail down the precise role of fiber itself, the practical message is clear: Eat a wide variety of fiber-rich whole foods.

Those foods should supply you with 20 to 35 grams of fiber each day. To get a rough idea of how much fiber you consume take the self-test on page 28. (To get a more precise tally of your fiber consumption, you'll need to consult the table on the facing page and read the labels on packaged foods.)

Untangling the Fiber Claims

The U.S. Food and Drug Administration now allows manufacturers to make certain claims about the fiber content and health benefits of packaged foods, provided the food is low in fat and sodium. Those claims can tell you at a glance whether the food supplies a significant amount of fiber:

- Products with the words "good source of fiber" or "contains fiber" must provide 2.5 to 4.9 grams of fiber per serving.
- Foods labeled "high" or "rich" in fiber must provide at least 5 grams.
- Products making a health claim about coronary disease or blood-cholesterol levels must contain lots of soluble fiber—more than 0.6

grams. That's particularly useful information, since labels generally do not differentiate between soluble and insoluble fiber.

Some bread manufacturers still try to mislead consumers about the fiber content and other nutritional benefits of their products by using deceptive names and colorings. Most breads called "wheat" rather than "whole wheat"—and even some multi-grain breads—are made primarily from refined flour, with just a smattering of whole-grain flour, plus some molasses or caramel coloring to give the bread that brown, "natural" look. Refined flour contains little fiber, since processing removes the outer, fiber-rich layer of bran.

If a package of bread—or cereal—carries none of the FDA-approved claims that would identify it as a significant fiber source, look for any of these reassurances on the label:
• The words "whole wheat" or "100 percent whole grain."
• Whole-wheat flour listed as the first ingredient. (If it's listed lower down, the bread contains less than 50 percent whole wheat, usually much less.)
• At least 2 grams of fiber in each slice of bread or each one-ounce serving of cereal.

How to Get More
If you eat the typical American diet, you'll need almost to double your daily fiber intake to get the recommended 20 to 35 grams. But don't go over-board: Getting a lot more than the recommended maximum can interfere with the body's absorption of nutrients such as calcium, iron, and zinc; hugh amounts of fiber can even cause intestinal blockage. And don't rush: Boosting fiber consumption too fast can cause gas, bloating, cramping, or diarrhea.

You're much less likely to encounter those problems if you get your fiber by eating fiber-rich whole foods rather than by taking lots of fiber pills or depending on high-fiber cereals. Even moderate reliance on fiber supplements rather than whole foods can deprive you of the many nutritional benefits of those foods. (Such supplements may be useful only for people who are constipated or who have trouble eating the grains, beans, and produce that supply fiber naturally.)

Eating at least five servings per day of fruits or vegetables and at least six servings of whole grains or beans—the amounts recommended by the U.S. Government—generally ensures an ample intake of fiber. Here are several ways to boost your intake of those high-fiber foods:
• Bake with whole-grain flour instead of refined flour. Choose brown rice over white rice, and whole wheat pasta over regular or even spinach pasta.
• Substitute whole, unpeeled fruits for fruit juices. Eat them as an appetizer, dessert, or snack. If you treat yourself to ice cream or cake, top it with fruit.
• Add fruit, brown rice, or whole-grain cereals to yogurt.
• Add beans, barley, or other whole grains to soups. Snack on cooked, cooled beans seasoned with garlic powder, chili powder, or Cajun spice. Or make them into a bean dip for raw vegetables.
• Prepare cold salads that combine cooked whole grains, pasta, or beans with chopped raw vegetables. (Don't depend on lettuce-based salads for fiber. Even the otherwise nutritious loose-leaf types of lettuce contain less fiber per serving than white bread does.)
• Make meatless entrees by cooking grains in seasoned stock, and tossing them with cooked vegetables or beans. ■

Questions

1. How does dietary soluble fiber decrease levels of cholesterol in the blood?

2. How does soluble fiber intake affect blood-sugar levels in diabetics?

3. How is insoluble fiber thought to decrease the risk of colon cancer?

Answers begin on page 161.

29 *The oils that many plants store in their seeds constitute a vast and renewable resource. The utility and importance of these plant lipids have been known for centuries. Today, plant oils are a source of chemical feedstocks for producing a wide variety of commercial products, ranging from soaps and detergents to plastics and lubricants. When research efforts directed at changing the genetic character of plants like the rapeseed plant, which produces canola oil, are achieved, plants will serve as a customized source of many valuable oils.*

New Oils for Old

Denis J. Murphy

Plant oils are a versatile source of renewable feedstocks for manufacturing a wide variety of industrial products, ranging from pharmaceuticals and cosmetics to lubricants and fine chemicals.

Following a relatively short-lived interruption in petroleum supplies during the oil crisis of 1973–74, Western economies suffered a pronounced decrease in industrial output, from which they took several years to recover. This event illustrates our dependance on petroleum, particularly as a source of fuels and petrochemicals. Petroleum has the advantage—for the present—of being available in large quantities at low cost, via (normally) reliable supply chains. Its main disadvantages include non-renewability, lack of biodegradability, requirements for extensive and often environmentally deleterious refining to make petrochemicals, and its contribution to net CO_2 emissions on combustion. The only renewable alternatives to petrochemicals are the so-called natural sources of oleochemicals such as vegetable oils, fish oils and animal fats or tallow.

Sources of Oleochemicals
Natural fats and oils have been used industrially since ancient times and were only supplanted in the late 19th century after the discovery of vast reserves of petroleum in the US and the Middle East. Since then, oleochemicals have remained a relatively minor but significant chemical feedstock, used in manufacturing surfactants, surface coatings and cosmetics. During the past few years, a number of scientific advances have coincided with economic and political developments in a way that has renewed interest in the expanding potential of oleochemicals and, in particular, vegetable oils.[1,2]

Improved methods of plant breeding and the increased use of fertilizers and biocides, together with considerable agricultural protectionism in the EU and US, have led to the accumulation of huge surpluses of food crops and the consequent removal from production, or 'set-aside', of 15–25 per cent of arable land. This set-aside land may provide an opportunity for cultivating greater quantities of subsidised oleochemical crops, such as rapeseed or linseed, for a decade or so into the future. Nevertheless, the longer term use of oleochemical crops is unlikely to depend on agricultural subsidies, which are subject to the caprices of political fashion, and are increasingly vulnerable since the endorsement by most industrialized nations[3] of the General Agreement of Tariffs and Trade (Gatt) in 1993. Instead, oleochemicals will eventually have to compete directly with petrochemicals at world market prices.

Basic Oleochemicals
Oleochemicals are mainly derived from triacyglycerols, often with relatively heterogeneous fatty acid compositions. The vast majority (78 per cent) of oleochemicals are derived from plants, and this trend should increase as the relative proportion of animal fats in total oleochemicals is predicted to decline from 22 to 18 per cent during the next 15 years. Almost 75 per cent of plant oils are derived from four major crops: soy, palm, rapeseed and sunflower. Of the 71 million tonnes (Mt) of annual plant oil production, only about 12 per cent is used for industrial oleochemicals; the remainder is for human or animal consumption. During the past decade, the centre of global plant

Reprinted with permission from *Chemistry in Britain*, April 1995.

oil production, and particularly of soy and palm oils, has been moving from the US/EU to developing countries in eastern Asia and South America. This is mainly because of the enormous expansion of palm oil production,[4] from only 1.1 Mt in 1970 to 11 Mt in 1990 and a projected 30 Mt by 2010.

In major plant oils, the C_{16} and C_{18} saturated and unsaturated fatty acids predominate. This tends to make these oils suitable for edible use, but of only limited value as oleochemicals. However, a number of surveys has revealed the enormous diversity of fatty acid compositions available in other oilseed species, most of which are not grown as crops. Plant scientists have found seed with high levels of individual fatty acids ranging in chain length[5] from C_8 to C_{24}. These fatty acids can contain up to three double bonds in *cis, trans* and conjugated conformations, or other functionalities such as hydroxy, epoxy and acetylenic groups.

Uses of Oleochemicals

Researchers are currently pursuing one of two main strategies to improve the fatty acid composition of oleochemicial feedstocks. First, they are using conventional and advanced breeding methods alongside genetic engineering to produce new varieties of important oilseed crops like rapeseed. Secondly, they can use modern genetic methods to accelerate the domestication of entirely new oilseed crops with high levels of particularly useful fatty acids.[6]

Plant-derived raw materials, usually cold-pressed or solvent-extracted triacylglycerols, are subject to a series of initial processing reactions to split fatty acids from glycerol and to produce the basic oleochemical feedstocks. Occasionally, the latter can be used directly to manufacture end products, but it is usually necessary to proceed via more complex oleochemical derivatives.[7] Oleochemicals are sometimes used in combination with petrochemicals, such as ethylene oxide and propylene oxide, in manufacturing end products.

The UK has an active oleochemical processing industry, although most of the raw materials are imported fro overseas, for example Malaysian palm oil and US soy oil. According to a recent survey, however, there are significant opportunities for UK-produced oleochemicals.[8] These potential uses fall into four broad categories: lubricants, surfactants, surface coatings and polymers.

Lubricants. Of the total UK lubricant market of *ca* 750,000 t, some 50–100,000 t (worth £20–40m) might be supplied by UK oleochemicals. These would have a particular competitive advantage in applications where the lubricant is lost to the environment, for example chain-saw lubrication, because of their superior biodegradability compared with traditional petrochemical-based lubricants.[8] In Germany and Scandinavia the use of oleochemical-derived lubricants is already mandatory in environmentally sensitive systems such as forests and inland waterways. Extending such legislation to the UK would clearly favour their use here.

Surfactants. UK-produced oleochemical surfactants cannot compete effectively with surfactants produced from coconut and palm kernel oil in Malaysia, Indonesia and the Philippines, unless we can genetically engineer UK rapeseed to produce over 90 per cent lauric (C_{12}) oil (normal rapeseed contains no lauric acid). This task is now well under way: in 1993, one of the first transgenic rapeseed varieties was produced by the US biotechnology company, Calgene. This contains a copy of a gene transferred from the California Bay tree, which produces seeds rich in lauric acid. The new transgenic rapeseed variety comprises over 30 per cent lauric acid, and has been released on a trial basis in California and Scotland. Scientists at Calgene are now attempting to increase the lauric acid content of rapeseed still further, by transferring additional genes from other species such as coconut.

Surface coatings. These can include paints, varnishes and inks. If we can produce the correct mix of fatty acid chain length and unsaturation in the new UK rapeseed varieties, they could compete with imported tall and soy oils as feedstocks for manufacturing the 80,000 t of alkyl paints produced annually in the UK. Petroleum-based printing inks contain carcinogens and are difficult to recycle because harmful solvents are needed to remove them from paper. On the other hand, oleochemical-based printing inks are non-toxic and can be removed from paper enzymatically. The increasing use of recycled paper and the willingness of consumers to pay a premium for 'environmentally friendly' products could lead to a demand for 25,000–50,000 t rapeseed oil for printing inks over the next few years.[8]

Polymers. Polymer manufacture consumes vast and increasing amounts of petrochemical feedstocks. In 1989, Pryde and Rothfus reported[9] that in modern cars polymers make up *ca* 25 per cent of bodyweight compared with just 7 per cent in the 1970s. They estimate that in the US alone some 22 Mt of polymer products was used in 1985. Oleochemicals currently make up a minor but growing proportion of polymer feedstocks. As well as their enhanced biodegradability, another advantage of oleochemicals is the pre-refining of the oils in the developing crop plant. For example, rapeseed oil has C_{18}–C_{22} fatty acids with double bonds at the C6, C9 and C13 positions, making possible high-grade polymer feedstocks like adipic (C_6), azelaic (C_9) and brassylic (C_{13}) acids without recourse to complex chemical engineering processes.

A far more futuristic venture is to produce the entire polymer molecule in the crop plant itself. This is the ambitious objective of a UK-based project that is under way at Zeneca in Berkshire, supported in part by Link funds from the Biotechnology and Biological Sciences Research Council (BBSRC). The aim of the project is to transfer several genes involved in polyhydroxybutyrate (PHB) biosynthesis from the bacterium *Alcaligenes eutrophys* to rapeseed, the seeds of which should accumulate high levels of PHB.[10] PHB is a biodegradable thermoplastic that is used for packaging, particularly of foods. The plastic can be broken down in a few months after being buried in the ground, but is completely stable under normal domestic storage conditions. PHB is marketed by Zeneca as Biopol and is currently produced by an expensive bacterial fermentation process. The production of PHB in rapeseed is likely to reduce costs and could considerably expand the market for Biopol—potentially requiring a further 100,000 ha of rapeseed to be grown in the UK.[8]

These are only a few of the many possibilities for expanding the use of plant-derived oleochemicals in the UK. Further efforts in this area will largely focus on manipulating the composition of storage oils in plants. This will require a greater understanding of the metabolic processes responsible for oil biosynthesis and how they are regulated.

Targeting Biosynthesis
The basic reactions of *de novo* fatty acid biosynthesis in plants are now known to differ in many respects from those of other multicellular (eukaryotic) organisms. For example, the dissociable multi-subunit fatty acid synthetase enzyme of plants resembles that of unicellular organisms (prokaryotes) like bacteria rather than the one- or two-subunit enzymes that are found in animals and fungi.[11] Last year, researchers also discovered that the enzyme responsible for the first step of fatty acid biosynthesis, acetyl-CoA carboxylase, occurs in two completely different forms in plants—one similar to that of bacteria and the other similar to the enzyme found in animals.[12,13] Despite these complexities, plant biochemists and molecular geneticists are gaining an increasingly detailed knowledge of the pathways of fatty acid biosynthesis in oilseeds. This information is helping them to clone ever-increasing numbers of the genes encoding the respective enzymes.[2]

The fatty acids of all organisms are assembled from acetyl-CoA via a fatty acid-synthetase enzyme complex. In most plants, the main product of this complex is C_{16} palmitic acid. This is elongated to C_{18} stearic acid, which can be progressively desaturated to oleic, linoleic and γ-linolenic acids. The latter make up the bulk of the oils that our major oil crops produce. These plants lack the enzymes responsible for producing more industrially useful fatty acids such as hydroxy or epoxy derivatives, which have actual or potential uses as plasticisers and coatings.[5] However, such enzymes do occur in other species, such as castor bean and *Vernonia*, which can accumulate large amounts of hydroxy or epoxy oils, respectively.

During the past five years, many of the genes involved in fatty acid biosynthesis and modification have been cloned, and techniques for transferring them between different plant (and animal or microbial) species have been developed. At present, the oilseed crop that can be most readily transformed and regenerated is rapeseed; soy transformation is possible, but less efficient. The widespread routine transformation of other major oilseed crops such as linseed and sunflower is still in its infancy. Transforming major perennial oil crops such as coconut palm and oil palm remains a rather long-term prospect, although preliminary efforts are under way in our own laboratories at the John Innes Centre in Norwich, and elsewhere.

Tailoring Oils
Over the next decade and beyond, we face the prospect of being able to engineer most major oil crops to produce the fatty acid composition of our choice. This presents plant biochemists and the oil-processing industries with new opportunities and challenges. These include the possibility of producing large (M t) quantities of relatively pure, semi-refined oleochemical feedstocks, such as C_8 fatty acids, C_{22} fatty acids of C_{18} epoxy fatty acids, at prices of $500–1000 t^{-1}. Alternatively, smaller scale production of higher value oils for niche markets is possible, eg γ-linolenic acid for therapeutic use, or C_{20}/C_{22} waxes for cosmetics. The challenge for chemists and engineers is to recognize these opportunities and to devise economic processes for converting these basic oleochemicals into useful products.

One example that illustrates this point is given by petroselinic acid, an isomer of C_{18} oleic acid with the double bond in the C6 rather than the C9 position. We can cleave petroselinic acid at its C6 double bond by oxidative ozonolysis to yield the C6 dicarboxylate adipic acid, plus C_{12} lauric acid. Adipic acid is an important bulk chemical feedstock for polymer manufacture, with an annual global market in excess of 2.2 Mt. The current process for manufacturing adipic acid from petroleum results in substantial emissions of the ozone-depleting greenhouse gas N_2O, so developing alternative, more environmentally friendly, routes is desirable. Petroselinic acid makes up 70–80 per sent of the seed oil of many of the *Umbelliferae* species, including the spice plant coriander. In laboratories in the US and UK, coriander genes are being transferred to rapeseed, with the aim of creating a new high-petroselinic oil crop.[14,15] An alternative and longer-term strategy that we are persuing at the John Innes Centre is to modify the coriander plant so that it can be grown as a high-yielding oilseed crop in the UK. The end result of these efforts should be a new oleochemical raw material with about 80 per cent petroselinic acid and priced at $400–500 t^{-1}. However, we will need to develop appropriate large-scale chemical and engineering methods for converting this oil into recognized industrial feedstocks, such as adipic and lauric acids, if it is to find a market.

A problem here is a lack of communication. Geneticists and biochemists are responsible for the specifications of oleochemical raw materials on the supply side, *ie* plant oils. Chemists and engineers, on the other hand, devise processes for using these oils to produce intermediate feedstocks (like adipic acid), and ultimately for manufacturing finished products such as plastics, paints and detergents. Unless we make efforts to bridge the academic divisions between these disciplines, opportunities to exploit new technologies may be missed. Given the current opportunities being opened up by biotechnology,[8] university and industrial research chemists should acquaint themselves with the range of fatty acids and their derivatives that is potentially available from plants. They could then experiment with substituting such molecules for existing petrochemical-derived products. Even more exciting is the prospect of developing novel products.

Keeping Good Council

The establishment in April 1994 of the Biotechnology and Biological Sciences Research Council (BBSRC), which covers a broad range of biology, chemistry and related engineering science, should catalyse this potentially fruitful dialogue. One of the principal aims of BBSRC is 'to advance knowledge and technology, and provide trained scientists and engineers, which meet the needs of users and beneficiaries (including the agriculture, bioprocessing, chemical, food, healthcare, pharmaceutical and other biotechnological-related industries), thereby contributing to the economic competitiveness of the United Kingdom and the quality of life'.[16] The interdisciplinary cross-fertilization in this type of endeavor is often the most productive part of such collaborations. Biotechnology has the potential to make available a huge range of basic raw materials, intermediate feedstocks and even final end products for the chemical industry. To take advantage of this opportunity, biologists, chemists and engineers must work together to identify and, if necessary, to manipulate those products and processes that are deemed worthy of future development.

Acknowledgements

I would like to thank Joanne Ross and David Fairbairn for their helpful comments during the preparation of this article.

References

1. D.J. Murphy, *Trends in Biotechnol.*, 1992, **10,** 84.
2. *Idem, Lipid Technol.*, 1994, **6,** 84.
3. K.D. Schumacher, *Inform.*, 1994, **5,** 408.
4. Anonymous, *ibid*, 1994, **5,** 715.
5. F.D. Gunstone, J.L. Harwood and F.B. Padley, *The lipid handbook*. London: Chapman and Hall, 1994.
6. *Designer oil crops,* D.J. Murphy (ed). Weinheim, Germany: VCH, 1994.
7. A.J. Kaufman and R.J. Ruebusch, *Inform.,* 1990, **1,** 1034.
8. D.J. Murphy *et al* in *Industrial markets for UK-produced oilseeds*. University of Reading: Centre for Agricultural Strategy, 1994.
9. E.H. Pryde and J.A. Rothfus in *Oil crops of the world*, G. Robbelen, R. Downey and A. Ashri (eds), p 87. New York: McGraw-Hill, 1989.
10. P.A. Fentem, *Can. J. Microbiology,* in press, 1995.
11. A.R. Slabas and T. Fawcett, *Plant Mol. Biol.,* 1992, **19,** 169.
12. C. Alban, P. Baldet and R. Douce, *Biochem. J.,* 1994, **300,** 557.
13. Y. Sasaki *et al, J. Biol. Chem.,* 1993, **268,** 25118.
14. J.B. Ohlrogge, *Plant Physiol.,* 1994, **104,** 821.
15. D.J. Murphy and D.J. Fairbairn, *World patent* WO9401565, 1993.
16. *Biotechnology and Biological Sciences Research Council (BBSRC) Corporate Plan 1994–1999*. Swindon: BBSRC, 1994. ■

Questions

1. From what sources are most oleochemicals derived?

2. When might oleochemical-derived lubricants be the best choice?

3. How would oleochemical-based printing inks be better than traditional petroleum-based inks?

Answers begin on page 161.

30 *Since 1940, the incidence of breast cancer has risen significantly, and sperm counts worldwide have fallen about 50 percent. Over that same period, the incidence of prostate cancer in some countries has doubled, and that of testicular cancer has tripled. Some scientists believe that these developments may share a common cause: the release of synthetic chemicals into the environment that mimic or block the action of the natural hormone estrogen. In humans, estrogen plays a range of developmental and reproductive roles in both males and females, some of which are only now beginning to be understood. Estrogenic substances in the environment may interfere with some of these processes.*

Estrogen: Friend or Foe?

Marilynn Larkin

"Sperm counts down worldwide." "Environmental estrogens linked to reproductive abnormalities and cancer."

These and other startling—and possibly misleading—headlines are based largely on results of recent research that appears to show an association between estrogen-like compounds in the environment to everything from "feminized" wildlife to reproductive problems in humans. While some scientists believe these findings are cause for concern, other researchers are convinced such fears are exaggerated.

"I know of no evidence that any industrial products are affecting human health," says Dan Sheehan, Ph.D., a research biologist in FDA's National Center for Toxicological Research. "There is some evidence for adverse effects to wildlife in highly polluted sites. But this hypothesis needs to be tested very carefully before any conclusions can be drawn."

Environmental estrogens come from many sources, Sheehan explains. One source is industrial products and byproducts, such as chemicals used to make plastic packaging, or those contained in pesticides. Estrogens and estrogen-like compounds may also be used as ingredients in cosmetics. Phytoestrogens are naturally occurring compounds in plants. And, of course, millions of women take estrogen pills for birth control and menopausal hormone replacement therapy.

Environmental estrogens, also known as "estrogen mimics" or "estrogen modulators," may be ingested, breathed in from the air, or absorbed through the skin. Some are considered "estrogenic"—meaning they imitate or enhance the effects of estrogen in the body; others are "anti-estrogenic"—they block or interfere with the body's use of estrogen. The estrogen-like compounds bind to estrogen receptors in cells, which may lead to changes in the cells, tissues, or organs.

Medical Uses of Estrogen

Millions of women increase the estrogen in their bodies when they take oral contraceptives or pills to relieve symptoms associated with menopause, such as hot flashes, sweating, and vaginal dryness. Certain estrogen drugs (Premarin, Ogen, and Estrace tablets, and Estraderm patch) also are approved to prevent osteoporosis. In addition, estrogens may be used to treat some forms of infertility and menstrual disorders.

Estrogen drugs may cause side effects such as bloating, weight gain, breast tenderness, and nausea. Women who take estrogen to treat menopausal symptoms should be closely monitored by a physician, and all women age 50 or older are advised to have regular mammograms, according to Enid Galliers, an FDA consumer safety officer.

The pills should be taken only when there is a "well-defined need," she says. "If a woman is not at high risk for osteoporosis and doesn't have unmanageable symptoms," she should carefully weigh the risks and benefits of estrogen in consultation with her health-care provider before making a decision about estrogen. She adds that estrogen drugs may increase the risk of uterine cancer in postmenopausal women who have not had a hysterectomy.

Do oral contraceptives promote or accelerate breast cancer? "The jury is still out," says Galliers.

Reprinted from *FDA Consumer*, April 1995, pp. 25–29.

"If woman has a family history of breast cancer, she might want to use another birth control method or be monitored closely by a physician."

Pregnant women should not take drugs containing estrogen because of the "potential harm of estrogen to the fetus, including a greater risk of birth defects in the reproductive system," she notes. It is this concern that is behind some scientists' belief that estrogen-like chemicals in the environment may have similar harmful effects, FDA's Sheehan adds. However, the effects of drugs that contain estrogen occur at therapeutic doses; levels of environmental estrogens may be much lower.

DES

"Because of what happened with [pregnant women taking] DES, we're concerned about the potential developmental toxicity of all estrogen-like chemicals, whether they are naturally occurring or industrial products," Sheehan says. "We must ask the question, 'Can this chemical have adverse effects on the fetus?' Right now, our database in this area is woefully inadequate. It is critical that more studies be done."

DES (diethylstilbestrol) is a drug that mimics a specific type of natural estrogen hormone called estradiol. DES, like other estrogen drugs, may be prescribed to treat advanced breast and prostate cancer, but is no longer recommended for use in healthy women. Potential side effects are the same as for other estrogen drugs.

In the 1950s and 1960s, DES was prescribed to millions of women in the United States and Europe to prevent miscarriage. This use was abandoned when it was learned that the children of women who took the drug during pregnancy have a high incidence of reproductive tract abnormalities and may be at increased risk for various types of cancer. Some researcher consider what happened to these children a model of what may happen to others who are exposed to estrogenic chemicals during intrauterine life.

Pesticides

Although prescription estrogen may produce side effects and increase a person's risk for cancer, as yet no studies have conclusively linked specific estrogen-like chemicals in the environment to problems in humans. However, there is some evidence that environmental estrogens cause reproductive problems in wildlife, Sheehan says.

For example, researchers discovered that a large percentage of eggs produced by alligators in Lake Apopka, in central Florida, failed to hatch or the hatchlings died within a couple of weeks. Moreover, alligators that survived produced minimal levels of estrogen, and thus were unlikely to be capable of reproducing. The culprit appears to be the estrogenic chemical DDE, a breakdown product of the pesticide DDT, according to a news report in the Jan. 31, 1994, issue of *Chemical & Engineering News*. High levels of DDT have been found in the alligators, probably as a result of chemical spills in the early 1980s when Lake Apopka was the site of a company that produced DDT-containing chemicals, according to the article. FDA scientists point out that the high levels of DDT might also reflect past use of DDT for mosquito control.

DDT also has been implicated in thinning and cracking of bald eagle eggshells, according to Ana Soto, M.D., an associate professor in the Department of Anatomy and Cellular Biology at Tufts University School of Medicine.

In an article in the October 1993 issue of the journal *Environmental Health Perspectives*, Soto and colleagues note that exposure to DDT—as well as other pesticides currently in widespread use—is associated with "abnormal thyroid function in birds and fish; decreased fertility in birds, fish, shellfish, and mammals; defeminization and masculinization of female fish, gastropods, and birds; and alteration of immune function in birds and mammals." However, it is not known whether humans experience similar effects from environmental exposure, and, if they did, what dosage level would be required to produce such effects.

Soto, who is also co-director of the Immunometric Assay/Protein Probe Core at Tufts' Center for Reproductive Research, and colleagues developed a test to screen for the estrogenic activity of chemicals in the environment. She testified at an Oct. 21, 1993, hearing before the U.S. House of Representatives' Subcommittee on Health and the Environment that test results showed the pesticides toxaphene and dieldrin (the use of which has been restricted in the United States by the Environmental Protection Agency) to be estrogenic.

A 1992 FDA Pesticide Program Report on Residue Monitoring during 1986-1991 concludes that any pesticide residues found in the foods sampled were within regulatory limits, meaning there is no cause for concern when it comes to human health. The report concludes that total dietary intake of pesticides during that period were "generally well below" the limits of acceptability set by the Environmental Protection Agency and the Joint (FAO/WHO Meeting on Pesticide Residues, panels of experts convened by the United Nations' Food and Agriculture Organization and the World Health Organization.

Industrial Products

Another group of chemicals that Soto's assay has shown to be estrogenic are antioxidants such as the nonylphenols. These chemicals "are shed by plastic products . . . and may contaminate foods during processing or packaging," she says. FDA is now conducting studies to determine the levels of nonylphenols in processed and packaged foods.

FDA also is studying another possible estrogen mimic called bisphenol-A. It is used in the production of polycarbonate plastics that sometimes are used to make water bottles and baby bottles. In addition, bisphenol-A is a component of the inner coating of food cans, according to Tom Brown who, before he retired from FDA in September 1994, was part of a team that investigated bisphenol-A in baby bottles. "We don't know if this is a real problem or not," he says. "We know that bisphenol-A is in the bottles; the question is, does it migrate out into the infant formula?"

If bisphenol-A is leaching into formula or other food products, scientists then have to determine the level of the residue and assess whether it might have harmful effects. The same is true of literally thousands of other industrial chemicals.

A news article on environmental estrogens in the July 15, 1994, issue of the journal *Science* noted, "As scientists choose sides in the increasingly sharp debate [about the potential health effects of environmental estrogens], there is one point of agreement: Chemicals that can potentially affect hormone levels are everywhere—in the food we eat, in the water we drink, in body fat, and in breast milk." Many of these chemicals occur naturally and so have been present in the environment for aeons. However, the question remains, are they harmful? We don't know yet.

Cosmetics

Estrogenic chemicals also are found in cosmetics. "Since we know at least 4,000 different ingredients are used in cosmetics, including synthetic chemicals and natural extracts from plant and animal sources, certainly there is potential for exposure to estrogen mimics from these products," says John Bailey, Ph.D., director of FDA's Office of Cosmetics and Colors. However, whether this should be an area of concern to consumers is "virtually unknown." More studies are needed but, he points out, this is not a simple task.

Unlike drugs, cosmetic products are not subject to premarket safety clearance, Bailey explains. "Cosmetic companies can use whatever ingredients they want to put together their products without providing data to FDA to substantiate safety. This doesn't mean these products are unsafe; however, they haven't undergone that extra level of scrutiny."

Although a cosmetic company is not required to present premarketing safety data to FDA, if adverse reactions occur after marketing, the company has to present FDA with data to back up the safety of the product.

"What is now being learned about estrogen mimics is of high interest and relevance. The science is unfolding," Bailey says. "However, we won't know whether there is real risk until these chemicals are studied in greater depth in all areas. Since the field is still very much evolving, we cannot make any recommendations at this time."

One area in which recommendations are somewhat more clearcut is in products containing hormone ingredients. In September 1993, FDA issued a final rule on topically applied hormone-containing drug products sold over the counter (OTC) and a proposed rule for cosmetic products containing hormone ingredients. The final rule established that any topically applied OTC hormone-containing drug (with the exception of hydrocortisone and hydrocortisone acetate for use as a topi-cal analgesic) is not safe and effective, and could no longer be sold OTC.

The proposed rule for cosmetic products containing hormone ingredients contains provisions to limit the amount of these hormones to levels that are "safe but not effective as drugs," Bailey explains. The source of the hormones (natural or synthetic) would also be listed on a product's label. Currently, the safety level of hormones in cosmetics has been established only for progesterone (concentration level up to 5 mg/oz) and pregnenolone acetate (concentration level up to 0.5 percent), in products labeled for use not to exceed 2 ounces a month.

FDA does not have enough information on estrogen hormones such as estradiol or estrone to determine if they can be safely used as cosmetic ingredients and at what concentration they no longer should be considered drugs. Therefore, the proposed rule would not permit the use of these estrogen hormones in cosmetics.

Phytoestrogens

One group of environmental estrogens with possibly positive effects is the phytoestrogens, which occur naturally in such foods as soybeans, whole wheat, and flax seeds. "I suspect that of all the environmental estrogens out there, the phytoestrogens are most likely to have human health effects," says FDA's Sheehan. These may include possible protection against certain types of cancer, including breast cancer.

Sheehan cites the example of Japanese women whose breast cancer rate increased significantly after moving to the United States and adopting a Western diet. "Some people believe that the higher fat level in the Western diet is responsible for the increase. But an alternative hypothesis is that the phytoestrogens in the typical Japanese diet may exert a protective effect. They may be anti-estrogenic and anticarcinogenic," Sheehan says.

Nevertheless, it would be "premature" to suggest that people modify their diets based on this hypothesis, Sheehan says. "We have to wait for results of extensive studies before any recommendations can be made."

Until results are obtained from in-depth studies that test the biological activity of environmental estrogens, as well as the potential effects of mixtures of these chemicals, there is no reason for people to become alarmed, Sheehan says. Such studies are being initiated in numerous laboratories around the country, including those at FDA.

"At this point, FDA is in a strong role with respect to raising these questions and investigating," he notes. "To a certain extent, we anticipated the problem, and resources were allocated to do this vital experimental work."

The outcome of such work may help provide guidelines for future use of estrogen-like chemicals, and determine whether the naturally occurring phytoestrogens have protective effects. Until then, the potential risks or benefits of these compounds for human health must be considered speculative. ■

Questions

1. List at least three reasons why women may take estrogen.

2. Why was DES prescribed to pregnant women, and what negative effect did it have on the children of the women who took it?

3. What type of government scrutiny do cosmetic products undergo before marketing?

Answers begin on page 161.

31 *The human brain consists of biomolecules organized into a highly sophisticated network that is able to perceive, think, calculate, self-repair, and feel. Computer designers may never be able to make machines that have all of the capabilities of a human brain, but many researchers feel that by using proteins, computer components can be built that significantly increase existing computer speed and memory.*

Protein Devices May Increase Computer Speed and Memory

Michael Freemantle

Prototype devices use bacterial protein memory cubes to store up to 300 times more information than current devices.

Researchers at Syracuse University have developed a prototype protein-based optoelectronic device that may provide as much as 300 times more memory than devices currently used in computers. The device uses laser beams to write information into and read information out of cubes containing the protein bacteriorhodopsin.

Bacteriorhodopsin is a light-harvesting protein found in a bacterium that grows in salt marshes. When the protein absorbs light, it undergoes a complex photocycle resulting in dramatic changes in its optical and electronic properties.

These unique photophysical properties were discovered in the early 1970s. Since then, bacteriorhodopsin's potential for use in optoelectronic devices has excited the interest of scientists around the world. A key breakthrough came last year when the Syracuse researchers discovered by chance a photochemical branching reaction in the protein's photocycle.

It is this branching reaction that provides for long-term, high-sensitivity volumetric data storage, explains chemist Robert R. Birge, director of the W. M. Keck Center for Molecular Electronics and research director of the New York State Center for Advanced Technology in Computer Applications & Software Engineering at Syracuse University.

Birge described the serendipitous breakthrough in a lecture on optoelectronic devices based on bacteriorhodopsin at a conference on "The Science of Complex Molecular Systems," held last month in Paris. The conference was *Nature*'s third international conference in Europe.

"One of the problems in making viable three-dimensional memories has always been that when you try to read or write information inside a volumetric solid, you cannot go back to the same location accurately, and you cannot rigorously exclude writing outside of the area that you are trying to write to," says Birge.

The Syracuse discovery overcomes this problem. "We use the protein as an optical AND gate," explains Birge. An AND gate is one of five main types of logic gates used in building integrated circuits. Binary-coded input signals are fed into the gate. The binary 0 state corresponds to "off," and the binary 1 state corresponds to "on." An AND gate yields a binary 1 output signal if and only if the two input signals are in the binary 1 state.

In their prototype device, a green laser beam is used to initiate the bacteriorhodopsin photocycle in an ultra-thin section of the cube. In effect, a "page" of molecules is photoactivated to initiate the photocycle. Two milliseconds later, before the molecules can relax back to their resting state, a two-dimensional array of red lasers at right angles to this beam is fired at the page. This induces the photochemical branching reaction to the "Q state" in the branched photocycle in sets of molecules on the page targeted by the red lasers. For computational purposes, these molecules are converted to the binary 1 state. Nontargeted molecules on the page return to the resting state along with all the other molecules in the cube. Computationally, these molecules are all in the binary 0 state. Both states remain stable for several years.

"Because you can move the page from one location to another inside the cube, you can rigorously select locations in the memory to write data," says Birge. The data are read by activating the page with the green beam again. Low-intensity red lasers are then fired at the page. Molecules in the binary 0 state absorb the light whereas those in the binary 1 state allow the light to pass through. The entire page of binary output signals is imaged by a charge-injection device array.

Data can be written in parallel at multiple locations on the illuminated page. Similarly, the multiple bits of data on a page can be read simultaneously. The simultaneous manipulation of multiple sets of data is known as parallel processing. Either sequence takes about 10 milliseconds, which is equivalent to writing or reading data at a rate of 10 megabytes per second for each page of memory.

This speed is comparable to a slow semiconductor memory, according to Birge. The overall speed of the memory is directly proportional to the number of cubes operating in parallel. An eight-cube memory would operate at 80 million characters per second.

Parallel processing is dependent on a "sequential one-photon" process, Birge says. The first photon, from the green laser, activates the photocycle, and the second photon, from the red laser, activates the branching reaction to form a stable photoproduct. "This sequential one-photon architecture completely eliminates unwanted photochemistry outside of the irradiated volume," he notes.

In principle, an optical three-dimensional memory can store approximately 1,000 times more information in the same size enclosure compared with a two-dimensional optical disc memory. Optical limitations and issues of reliability lower this ratio, however. "A three-dimensional memory stores roughly 300 times more information in the same volumetric enclosure," Birge explains. "This means you can make a memory that would store about three and a half gigabytes [3.5 billion bytes] on a card that would fit into a typical computer."

The Syracuse team has benchtop models of the bioelectronic device operating in its laboratory. These fall into the category of level I prototypes. Level II prototypes are miniaturized versions that can be put into a computer. "My prediction is that we will have working level II prototypes in three to four years," says Birge.

The devices will need further improvement before they reach the level III prototype stage. At this stage, explains Birge, the devices will be ready to manufacture. "The field is moving very quickly, and it's well funded," he says. Birge's research is sponsored by the U.S. Air Force Rome Laboratory, the W. M. Keck Foundation, the National Institutes of Health, and 12 companies including four major U.S. computer manufacturers.

The aim is to make a memory card with the protein encapsulated in a polymer. "The protein is very cheap for storing data compared with semiconductors," comments Birge. He notes that, at current prices, disks cost approximately $1.00 per megabyte of memory. "Our goal is to make a 3.5 gigabyte card which would sell for about $600. That is a considerable savings, although I'm sure prices of disk drives will drop. However, I anticipate our card will be cost-effective to the user."

One of the major attractions of using the bacteriorhodopsin in computer memories is its thermal and photochemical stability. The protein occurs in a lipid matrix in the purple membrane of a bacterium that grows in salt marshes where the salt concentration is around six times that of seawater.

The bacterium, called *Halobacterium halobium,* is able to survive in the harsh conditions of high temperature and intense light in the salt marsh by switching from respiration to photosynthesis when the need arises.

The bacterium grows the purple membrane when oxygen concentrations become too low to sustain respiration. It absorbs energy from light by means of a chromophore embedded in the protein. When the chromophore absorbs light, the structure of the protein changes and the bacterium is thus able to pump a proton across the membrane. This gives rise to a chemical and osmotic potential that serves as an alternative source of energy.

Bacteriorhodopsin can undergo more than 1 million photochemical cycles without damage. According to Birge, this value is much higher than most synthetic photochromic materials. And, he notes, "Few semiconductor devices could survive a salt marsh."

The Syracuse researchers grow the bacteria in simulated salt marsh conditions in their laboratory. "Isolating the protein is essentially straightforward," comments Birge. The team uses genetically modified bacteria that produce 10 times as much of the protein as normal, he says. "We can create about 500 mg a week per 14-L vessel. That's a lot of material."

Chemistry professor George M. Whitesides of Harvard University drew attention to the stability of the protein in his summary of the Paris conference. "Bacteriorhodopsin, in the hands of Birge, has proved to be much more robust and has more interesting properties than I think many people would ever have guessed," he comments. "It is a real example of how you can use proteins for something that requires real stability."

Birge's lecture at the Paris conference was one of four lectures presented in a section on molecular electronics. Molecular electronics, according to Birge, can be broadly defined as the encoding, manipulation, and retrieval of information at a molecular or macromolecular level. Several researchers commented on the limitations of semiconductor technology and the potential advantages of using molecular electronics in computers.

Allen J. Bard, Ackerman/Welch Regents Chair in Chemistry at the University of Texas, Austin, tells C&EN: "I think, in silicon technology as practiced now, one will reach a stopping point based on the material capabilities. You can only do so much with lithography and with small wires."

Michael C. Petty, codirector of the University of Durham Centre for Molecular Electronics in England, reinforces the point. "There's a lot of interest in materials other than silicon and gallium arsenide for electronics because these are somewhat limited in their applications," he tells C&EN. "I don't think we're tying to replace the silicon chip. But there are lots of areas where organic materials might fulfill a specific role. What Birge is doing is extremely exciting. Clearly, he's bringing together expertise from the background of physics, chemistry, and biology."

Birge stresses that the development of a purely biomolecular computer is not a serious proposition. He sees the development of a hybrid technology based on a combination of biomolecules and semiconductors as far more likely. "Liquid crystal display technology is a prime example of a hybrid molecular-semiconductor technology that has achieved widespread success," he says.

One of the problems that has to be solved with protein memory cubes is the provision of an environment for the protein that will give it stability over decades rather than a few years. "Right now, our memories will store information for five years plus or minus a year without any trouble," says Birge. "It's going to take a lot of additional work not only using chemistry but also using molecular biology to make something that can store data on the order of 20 years."

Birge considers protein memory cubes to be more environmentally friendly than semiconductors, however. "There are several steps in making a semiconductor device, such as cleansing and etching, that involve a major amount of pollution," observes Birge. "The preparation and isolation of the protein is environmentally safe, and you can even eat the protein. It is nutritious, but not delicious."

The Syracuse team is also working on other types of memory applications for bacteriorhodopsin. One of these is the use of the protein to build neural networks that mimic the learning-by-association capabilities of the brain. This type of memory is known as associative. An associative memory receives an image, or block of data, and then searches the entire memory for a data block that matches the input. The memory then returns the matching or most closely matching block as output.

"You feed in a small amount of data and the memory finds that location in memory that has the best overlap with the data you fed in. It then feeds back not only that piece of information but all the associated information," explains Birge.

Birge's laboratory has developed an associative-memory device that relies on the holographic properties of thin films of bacteriorhodopsin. "The device takes advantage of the fact that when the protein is activated with light, it changes its refractive index dramatically," he says. "During this refractive index change, it can function as a holographic recording material." Holograms allow multiple images to be stored in the same memory segment. This enables large sets of data to be analyzed simultaneously. ∎

Questions

1. What protein is involved in these protein memory devices?

2. In principle, how much more information can be stored in an optical three-dimensional memory as compared to a two-dimensional optical disc memory?

3. Why are protein memory cubes more environmentally friendly than semiconductors?

Answers begin on page 161.

32 *Proteins don't get a lot of press, at least not compared to DNA, the biomolecules holding the very blueprints of life. Yet proteins are indispensable components of all living things. They play crucial roles in biological processes. Proteins are the workers (machines) of the cell, assembling, modifying, and maintaining various structures. Recently, chemists have reported some success at protein design and synthesis in an approach to making useful biomolecules.*

Designer Proteins: Building Machines of Life from Scratch

Richard Lipkin

Imagine that engines, like mushrooms, grew out of the ground.

The ancients wandering amid their fields might have stumbled upon strange iron hulks sprouting from the soil. Examining the metal contraptions, they might eventually have figured out that the intricate machines can do work, that their power can be harnessed, that they can generate electricity and move vehicles.

These early investigators probably would have taken the engines apart, determined how they were made, and learned how to operate them. Ultimately, they would have put their knowledge to the test by building one from scratch.

For today's molecular biologists, proteins offer this same allure and challenge. Often called the basic machines of life's biochemical factory, proteins carry out an extraordinary array of cellular functions and provide much of life's physical structure. Therefore, to master the intricacies of protein chemistry one must delve into life's fundamental mechanisms.

"Given that proteins are the machines of life, then to design proteins from scratch is like doing nanotechnology on the biological front," says Michael H. Hecht, a chemist at Princeton University. "To design and build a protein de novo is the ultimate test of our understanding of how proteins work. If we can learn to build these molecular machines, make them to order, and tailor them to our specifications, then we don't have to rely only on the [proteins] that nature provides."

The ability to tailor proteins to carry out very specific chemical tasks, thus harnessing their capacity to do molecular work, offers tremendous potential. Be it in medicine, industry, or environmental remediation, the customized protein beckons to biochemists as the steam engine once did to the earliest industrialists.

Recently, chemists have been reporting incremental successes at protein design, which are being hailed as milestones in the long march toward making useful biomolecules.

Among the latest achievements is the synthesis of beta-sheet proteins. Much harder to make than their alpha-helical cousins—which are long, winding strands—the beta-sheet proteins have been likened to "a greasy sandwich," with one side of each molecular sheet consisting of a slippery hydrophobic, or water-repelling, surface. Like pieces of bread spread with peanut butter, the two sheets come together—and stick.

Whereas a single alpha-helix consists of an elegant spiral of 15 to 20 amino acids, the beta sheet requires 60 to 70 amino acids. An alpha-helix protein will take shape in solution, yielding a broth of single, floating strands. Beta sheets, in contrast, are much less manageable. They form only as evenly matched pairs pressed together as a sandwich. And they must match face-to-face in just the right position.

"An alpha helix is like one hand clapping. It doesn't make much sound, but it can exist," says Bruce W. Erickson, a chemist at the University of North Carolina (UNC) at Chapel Hill. "But a beta sheet cannot exist alone. It must rest against another hydrophobic structure, most notably another

Reprinted with permission from *Science News*, the weekly newsmagazine of science, copyright 1994 by Science Service, Inc.

beta sheet. And it's quite difficult to get them to fit together in just the right orientation."

"Imagine trying to make a sandwich by slapping two pieces of bread together from 2 feet away," he observes, "then trying to slide them into place so that the sandwich looks well made." Rather, he wants to design a beta-sheet protein whose two sandwich halves can easily find each other and fall readily into place.

For the engineered molecule to survive as a working protein and not just collapse into a garbled mess, it must fold into a specific shape. Tweaking the peptide's three-dimensional geometry into place presents one of the toughest tasks of protein design and synthesis. A successful structure depends entirely on a successful fold.

In a sense, designing an amino acid sequence that will fold into a properly shaped protein is like creasing a piece of origami paper so that, when immersed in a solution, the water's energy will fold it into a swan.

"Proteins don't really fold themselves," says Erickson. "Water folds proteins."

While some pieces of a protein molecule attract water, others are repelled by it. The art of protein design comes in putting just the right pieces in just the right places so that, merely by reacting to its watery environment, the complex chain of amino acids will collapse into exactly the right three-dimensional sculpture.

Much of a protein's crumpled form comes from the efforts of its water-repelling side chains to escape its aqueous surroundings and squeeze into the protein's protected core. To design a protein that will put itself through such contortions and then build it amino acid by amino acid requires extraordinary precision, physical intuition, and years of trial and error.

Hundreds of proteins may fail before one gives even a hint of success.

Erickson's journey into protein design began 12 years ago, when, at a conference in France on gene sequences, he and Jane S. Richardson, a biochemist at Duke University in Durham, N.C., conceived of a way to build a novel bell-shaped protein from scratch. Back in the United States, synthesis of the first such betabellin proteins began.

"The first eight molecules, which took 5 years, were failures," says Erickson. "Then betabellin 9 began to work." Refining, testing, and redesigning have taken his group up to betabellin 17.

Now Erickson, working with Yibing Yan, also a chemist at UNC, is reporting details of the design, synthesis, and characterization of several variations on the betabellin theme. One of them, betabellin 15, appears to demonstrate activity similar to that of a naturally occurring enzyme.

"We're attempting to redesign betabellin 15, using computer-assisted models, to enhance the enzymatic activity by repacking its interior so the hydrophobic residues fit together better. These are very exciting results to get after 12 years of work," says Erickson, whose most recently published findings, on beta-bellin 14D, appeared in the July PROTEIN SCIENCE.

"A long line of dedicated people has struggled to solve the protein's inherent structural problems," he adds, highlighting the painstaking nature of adding a bond here, a crosslink there, then purifying and analyzing the compound to see if the design worked.

"Ten years ago, we'd have given our eyeteeth to be half as far as we are today," Erickson observes.

Taking a slightly different approach to the beta-sheet conundrum, biochemist Thomas P. Quinn of the University of Missouri in Columbia has reported successfully designing and synthesizing from scratch a protein he calls a "betadoublet."

Itself a sandwich-style protein composed of two sheets pressed together, the betadoublet uses specially placed disulfide bonds as glue. Along with Duke's Jane Richardson and other colleagues, Quinn described the synthesis in the Sept. 13 PROCEEDINGS OF THE NATIONAL ACADEMY OF SCIENCES.

"This motif is found in many naturally occurring proteins, such as immunoglobulin," says Quinn. The overall goal, he adds, is to use "design principles to make and evaluate a stable, three-dimensional beta-sandwich protein, using only naturally occurring amino acids." The end product should ideally resemble a native, or naturally occurring, protein.

For inspiration, Quinn looked to the betabellin series of proteins, but he then decided to branch off in a different direction. Instead of building the molecule piecemeal, he created a synthetic gene to manufacture beta sheets and spliced it into *Escherichia coli* bacteria. Once the bacteria expressed the gene, they cranked out beta sheets. The group then purified the proteins and searched for ones with the desired three-dimensional structures.

"There's a big hurdle we have to get over," says Quinn, "and that is to get the designed protein to have a stable, unique core." In naturally occurring proteins, dangling "side chains" normally fold neatly into place within the molecule's interior. Such a careful collapse, he says, is essential to its structure.

The trouble with designed proteins, Quinn points out, is that they "usually adopt the correct secondary structures and condense into compact molecules." Unfortunately, those critical side chains "often flop around, which means the core isn't quite right."

"We can get 80 percent of the correct structure, but then the core gives us trouble," says Quinn. Currently, he and his research group are attempting to redesign the beta-doublet's interior so that the core locks more firmly into place.

"The core is really critical," he stresses. "A protein's structure depends on the core. It's the last piece of puzzle. If we can get over this hurdle, we're home free."

Meanwhile, Hecht and his colleagues at Princeton are pursuing what they call a general strategy for de novo protein design. For a protein to fold properly, they hypothesize, its amino acid sequence must contain hydrophilic, or water-loving, and hydrophobic sections in very specific locations. Less important, they stress, are the details of which amino acid goes where and in what order. Less worrisome too, they maintain, are certain details of the protein's side chains in determining how the chains collapse into the molecule's core.

Instead, Hecht's team highlights the necessity of having water-attracting and water-repelling cornerstones in just the right spots along the protein's amino acid sequence, leaving the rest of the structure to take care of itself.

"The exact packing of the protein's three-dimensional jigsaw puzzle does not have to be set in advance," says Hecht. "You only have to specify the sequence location, not the identity, of the hydrophilic and hydrophobic residues."

The theory thus provides a recipe for assembling a protein that will fold in a specific way.

To test the hypothesis, Hecht and his coworkers engineered 48 genes specially designed to make proteins that would fold into a bundle of four helices. The proteins that these genes encoded all had one thing in common: the same pattern of water-loving and water-repelling units in the same key spots. In the end, 29 of those genes produced stable, compact proteins that folded more or less as predicted.

"The take-home message here is that the protein took care of itself," Hecht says. "We just had to provide the driving force of basic structure and hydrophobic collapse. Then everything else fell into place."

While Hecht's general strategy so far is geared to produce a four-helix bundle, this type of approach may prove useful later on for designing beta-sheet structures or other varieties of biomolecules.

Indeed, David A. Tirrell, a polymer scientist at the University of Massachusetts at Amherst, has such visions in mind. In the Sept. 2 SCIENCE, he and his colleagues describe techniques for exploring the control of "supramolecular organization" in genetically engineered polymers.

Such polymers, he says, could create an entirely new type of material.

"Appropriately designed artificial proteins," Tirrell says, "represent a new class of macromolecular materials, with properties potentially quite different from those of the synthetic polymers currently available and in widespread use." For instance, from a carefully designed artificial gene sequence, Tirrell has successfully synthesized a layered crystal made up of stacked beta sheets.

"Trying to build proteins is a very humbling experience," Quinn says. "The alternative approach is to take a native protein, tinker with its structure, and try to modify it. That's a valid approach, but tough, too. Over millions of years, evolution has honed each protein to achieve a stable shape to do a specific job. You can alter a few amino acids, but most of the time that has little effect because of the molecule's evolutionary stability.

"If you want to learn how proteins fold in general, just from the information encoded in their amino acid sequences, then you have to free yourself from some of the confines imposed by evolution. And the only way to do that, really, is to build one from scratch and see what happens."

"It's like learning a new code of nature with a new set of rules," Quinn adds. "The DNA code only has four bases in a linear sequence. Figuring that out took a long time, but look what it made possible. With proteins, there are 20 components in three dimensions"—that is, configurations of the 20 naturally occurring amino acids.

Indeed, deciphering this chemical architecture may come about as much from building up proverbial protein sand castles as from tearing them down—the traditional process of analysis and synthesis.

"Nature has been at this design game for a long time," Erickson notes. "Here we are, trying to do in a few years and a few labs what nature has been doing for 2 billion years—namely, running an open-ended experiment with billions of living laboratories, each one an individual whose life depends on its enzymes.

"This is nature's protein-engineering project," he adds. "We call it life." ∎

Questions

1. What determines protein shape in water?

2. What holds a "betadoublet" together?

3. What hypothesis is the basis for the Princeton group's strategy for de novo protein synthesis?

Answers begin on page 161.

33 *Enzymes are the most efficient of all known catalysts; some are able to increase reaction rates by a factor of 10^{20} times that of uncatalyzed reactions. The use of enzymes as industrial catalysts is undergoing rapid development due to the low energy requirement and cost efficiency of enzyme-catalyzed reactions. Because enzymes are biodegradable, Denmark's chemical company, Novo Nordisk, is developing a number of enzyme products as environment-friendly alternatives to traditional industrial chemicals. The company has become a pioneer in "green chemistry," searching the world for enzyme substitutes for synthetic chemicals.*

Novo Nordisk's Mean Green Machine

Julia Flynn, with Zachary Schiller, John Carey, and Ruth Coxeter

It's at the top of a growing market: Finding natural substances to replace chemicals.

When scientists at Denmark's Novo Nordisk—and even their families—go on vacation, they pack more than their swimsuits: They fill their suitcases with soil-collection kits to gather exotic, enzyme-producing microbes. It pays off. In the soil of an Indonesian Monkey Temple, the father of a Novo scientist unearthed an enzyme that is now widely used by soft-drink suppliers to change starch into sugar. In a pile of leaves at a Copenhagen cemetery, a researcher picked up a bug that produces an enzyme that can be used in detergents to help remove protein stains. "We're tying to find natural solutions to industrial problems," explains Lisbeth Anker, who heads Novo's worldwide microbe search.

Screening thousands of soil samples to find a single potential product is a bit like scouring the beach for a lost contact lens. But, with help from biotechnology, Novo has developed more than 40 industrial enzymes for everything from stonewashing jeans to ripening apples more quickly. In the process, the $2 billion company has emerged as a pioneer in "green chemistry"—finding more benign substitutes for synthetic chemicals. Decades before many rivals, Novo saw that synthetic chemicals would be attacked for harming the environment, and that enzymes could perform many of the tasks chemicals traditionally do—acting as solvents or flavor enhancers, for instance—with less cost to the environment. Its head start and innovative products have helped Novo grab a 50%-plus share of the $1 billion world enzyme market. Now, by tonnage, Novo produces more biotech products than anyone else—including its mainstay pharmaceutical products, insulin and human growth hormone.

Today, the rest of the chemical industry is trying to catch up. Biologically based chemicals still make up just a sliver of the overall chemical market. But the public's chemophobia is escalating, while regulators continue to clamp down on which chemicals industry can use and how. These forces, plus a heightened interest in "sustainable development"—the idea that future economic health depends on cutting pollution and using fewer natural resources—are fueling a conversion to greener alternatives by chemical makers such as DuPont, Dow, and South San Francisco's Genencor International. "Products that support sustainable development will be in demand," insists Mads Ovlisen, Novo's chief executive. "I see no alternatives."

Kicking Habits
Fact is, enzymes are a necessary alternative for Novo, too. Margins for its mainstay drug, insulin, which commands a 44% share of the world market, have fallen since the drug went off patent decades ago. And now a shortage of insulin caused by a delay in the startup of a

Reprinted from the November 14, 1994, issue of *Business Week* by special permission. Copyright 1994 by The McGraw-Hill Companies.

Novo plant in North Carolina is threatening to drive customers to Novo's chief insulin rival, Eli Lilly & Co. Investors are worried enough that Novo stock is selling at a discount to the Danish market. For all those reasons, Ovlisen wants to boost enzyme sales and make Novo less dependent on insulin.

To do that, the company is investing heavily in enzymes, which account for $575 million of its sales. This year, it spent $120 million to expand Novo Nordisk BioChem Inc., its operation in Franklinton, N.C. And in May, Novo acquired a site in China for a plant to supply enzymes to China's growing detergent, brewing, and textile industries—an investment that could reach $210 million in the next decade. At the same time, Novo is spending 10% of enzyme sales on research and development to keep rolling out innovative products.

Novo didn't set out to be a pioneer. The enzyme business began as a sideline to the pharmaceutical business, which still accounts for more than 66% of overall sales. For decades, Novo produced penicillin through fermentation—a process of growing microorganisms in an organic soup from which the enzymes the bugs make are extracted and turned into drugs. In the early 1960s, Novo scientists started experimenting with other enzymes after a Swiss company, Gerbrüder Schnyder, developed the first detergent enzyme. Novo quickly followed with its own product, Alcalase, a protein-stain remover that bested Schnyder's in alkaline washing conditions. Within five years, Proctor & Gamble, Unilever, and Colgate-Palmolive had added it to detergents. Buoyed by that success, Novo rapidly expanded into industrial enzymes.

The business, which is growing by 10% to 15% a year and commands premium prices for novel products, is a promising area for future growth. Enzymes—natural catalysts that can speed up a chemical reaction without being consumed in the process—are biodegradable. And since they work best in mild conditions, they can require up to a third less energy to use than many synthetic chemicals do. That helps explain why last year Novo sold $230 million worth of enzymes to the $25 billion detergent industry—an 8% gain—even as U.S. soap sales fell because of price cuts.

Novo's detergent additives are also in demand because they work better. In collaboration with Novo, Procter & Gamble Co. developed Carezyme—a proprietary enzyme that reduces the fuzz buildup on fabrics. Last year, P&G added Carezyme to its worldwide stable of detergents. It now boasts of the enzyme's benefits with its Tide slogan: "Keeps clothes looking more like new." The Cincinnati giant cites Carezyme as one reason brands such as Cheer outperformed the market last year.

There are plenty of other instances where Novo's creativity has paid off. "You don't get and keep a 50% market share without a constant stream of innovative products," says Paul Krikler, an analyst at Goldman, Sachs & Co. in London. Take hexane, a chemical solvent used to extract oil from linseeds and other vegetables. Since it is derived from crude oil, hexane is explosive and dangerous to breathe. Novo's solutions? An enzyme that extracts the oil from linseeds that are doused in water, which cuts the risks to workers and to the environment.

Similarly, the pulp and paper industry is coming under fire for its use of chlorine to remove residual lignin, a part of the wood that will turn yellow or brown if it's left in paper. Novo sells an enzyme that makes lignin easier to remove—cutting chlorine use by about 30%. It expects to have an enzyme to replace chlorine altogether within two years. The market for such an innovation could be worth hundreds of millions of dollars. "We're betting that industries like paper and pulp will be willing to pay for solutions to their environmental problems," says Ovlisen.

The 54-year-old Ovlisen is credited with quickly grasping, when he became president and CEO in 1981, how Novo's long-standing fermentation expertise could be harnessed to biotechnology. Biotech enables researchers to move genes from one type of cell to another: Novo's fat-splitting detergent-additive enzyme, Lipolase, for instance, was found in a strain of fungus that produced it only in minute quantities, making industrial production difficult. But Novo scientists transferred the gene to another microbe able to churn out more of the stuff.

On the Lookout
Since most genes can be transferred to a more productive host, Novo has stepped up its hunt for novel microbes that will do everything from taking fat out of foods and mopping up toxic waste to improving the digestion of farm animals. Last year, it created a new, 30-person R&D group in Davis, Calif., that is charged with finding new enzymes in extreme environments, such as the boiling springs of Yellowstone National Park, the deep oceans, or the frozen reaches of Antarctica. That's in addition to the 550 researchers who are already working in Denmark on enzymes and biopesticides. All researchers are encouraged to experiment with new ideas and can spend up to 20% of company time on their own projects.

Today, Novo has a solid roster of products in the pipeline. In the next few years, the company plans to roll out enzymes that do everything from chewing up toxic chemicals and killing insects harmful to plants to making industrial cleaning jobs easier. Increasingly, those enzymes will be designer creations—their structures modified by genetic engineering so that they are precisely tailored to do a particular task or to operate at a specific temperature.

Novo has paid a price for its technological leadership. In the late 1960s, for instance, reports that workers in detergent plants were suffering allergic reactions fueled public fears that enzymes were unsafe. In response, P&D and other manufacturers yanked enzymes from their U.S. products. Novo enzyme sales, the biggest in the industry, dropped 30%, from $86 million in 1969 to $60 million in 1970, forcing the company to lay off 400 of its 2,100 workers. German and Japanese rivals dropped out of the business, but Novo hung in—eventually figuring out how to granulate the enzymes to solve the allergy problem. Yet even today Novo is under attack by European opponents of genetic engineering, who fear that the technology cannot be adequately controlled.

Those critics, however, haven't prevented Ovlisen from plowing ahead in the use of biotechnology for Novo's drug and enzyme businesses. Indeed, it spends more on R&D than do its chief rivals, Gist-Brocades of the Netherlands and Genencor International in South San Francisco. Partly as a result, analysts estimate, Novo will earn $399 million before taxes in 1995, up 15%, on sales of $2.7 billion.

Remaining an innovator is critical. Enzymes have a short life cycle, so Novo needs to keep churning out novel replacements quickly. Meantime, rivals such as Genencor, a joint venture of Eastman Kodak, Finland's Cultor, and Genencor Inc., are becoming more formidable competitors. Genencor, which had $150 million in sales last year, just rolled out a stonewashing enzyme and is racing Novo to market with its own chlorine substitute. Novo's challenge will be to keep pumping out novel products while its cash-cow insulin business, which has financed its investments in enzymes, slows down. Under the circumstances, it's no wonder that Novo scientists are stepping up their microbe-hunting safaris. ∎

Questions

1. What ingredient in Tide detergent allows the company to boast, "Keeps clothes looking more like new," and what does this ingredient do?

2. Why is hexane dangerous, and why is Novo's enzyme better?

3. How did Novo overcome the difficulty of industrial production of Lipolase?

Answers begin on page 161.

34 *Enzymes have been of major importance since ancient times, when these proteins were first used in the production of wine and cheese. Today, enzymes play vital roles in a number of food-processing and pharmaceutical industries. To expand the range of processes in which catalysts might be used, chemists are exploring the development of enzyme mimics: chemically synthesized enzymelike structures. Another approach involves protein tailoring, making specific chemical modifications to existing enzymes so that they have a new catalytic role.*

Better by Design: Biocatalysts for the Future

Alan Wiseman

Useful industrial biocatalysts are currently being redesigned by using biological and chemical techniques, to provide novel enzymes and mimics for use in food processing and pharmaceutical manufacture.

For optimum performance, industrial enzymes (process biocatalysts) must be compatible with the desired operating conditions: temperature, pH, non-aqueous steps, reactor configuration and size, and with other enzymes in the process. The required enzymes can be made by genetic engineering, as a result of introducing specific desirable gene mutations. In the future we may also be using catalytic antibodies (abzymes)—now made by immunising animals with the corresponding transition state analogues.

Protein tailoring, which involves making specific chemical modifications to the isolated enzyme, often provides an alternative approach, as does the use of biochemically or chemically synthesised enzyme-like synzymes or enzyme mimics. Because the full spatial conformations of most enzymes are uncertain, attempts to design new enzymes are somewhat empirical, being partly based on computer predictions of useful amino acid sequences, by analogy with known structures determined by high resolution X-ray crystallography.

Redesign and Mimicry

Bioconversions using natural enzymes usually exhibit higher reaction rates than those achieved with chemically prepared enzyme mimics at moderate temperatures and roughly neutral pHs. Mimics are designed to bind the substrate, and then to catalyse the passage of the substrate through the transition state complex, to form products.[1] Enzymes and mimics are able to lower the binding energy of the transition state, thereby reducing the activation energy for substrate conversion.

In food or pharmaceutical applications, it is important to ensure that the enzymes or mimics are non-toxic. One solution may be to bind the enzymes covalently on an immobilised support, preventing them from contaminating the product. Surprisingly, however, despite over 25 years' experience of using immobilised enzymes, process technologists have been reluctant to adopt such procedures in industrial processing.[2]

The applications of enzymes in food processing have been delayed because it is expensive to produce enzymes pure enough to obtain the appropriate specificity, efficacy and safety. Enzymes are sold on an activity basis, not by weight, however, and high purity is unnecessary if they are needed for hydrolysing macromolecules such as polysaccharides and proteins.[3] Such hydrolytic reactions are usually beneficial in food processing, where they are associated with fewer problems in pumping, mixing, stirring, and filtration, and the subsequent energy savings. In addition, we use enzymes to improve the taste, texture and composition of the products, and occasionally to reduce their toxicity. Most natural enzymes, except those from bacterial sources, are 'generally recognised as safe' and therefore little benefit would be derived from using a more costly, redesigned hydrolytic enzyme.

Reprinted with permission from *Chemistry in Britain*, July 1994.

However, redesigned hydrolases may be useful when natural enzyme mixtures cannot be easily separated, for example hemicellulases and cellulases, to discriminate between the respective substrates.

Future industrial processes will require designer enzymes with a range of activities, but it would be naive to consider hydrolases as simply first generation 'simple-mechanism enzymes' that will be easy to improve. For example, although the intestinal protease (hydrolase) chymotrypsin has no cofactor,* its activity is nevertheless based on an exact juxtaposition of particular amino acid residues in the hydrophobic core of the globular protein molecule. During catalysis, a buried aspartate residue initiates a charge-relay from amino acids Asp102 and His57 to the Ser195 at the surface—a hydrophilic area that interacts with water. The aspartic acid residue buried in the core of the enzyme has an unionised carboxylic acid group while the serine hydroxyl group at the active site is ionised by losing a hydrogen ion. Only the negatively charged oxygen of the serine residue is able to form a covalent bond with the carbonyl of the peptide bond adjacent to the carbon that links to the side chain of the aromatic amino acid residue (phenylalanine, tyrosine or tryptophan). A water molecule can then attack that peptide bond and re-form the carboxylic acid and amino group respectively, to achieve hydrolytic proteolysis at that position in the polypeptide chain of the protein molecule. We should be able to improve these serine proteases by redesign strategies including protein engineering or protein tailoring. For example, converting Gly196 to alanine increases the thermal stability of the serine protease subtilisin, which is responsible for the biological action of washing powders.

*Generally, an organic moiety or metal ion, the presence of which is needed to activate the enzyme.

Chymotrypsin can also hydrolyse model esters by this serine-dependent mechanism, and some success has been achieved with enzyme mimics, although reaction rates are lower.[4] Nevertheless it is difficult to mimic the full specificity of chymotrypsin. Chymotrypsin itself can be thermally stabilised (100-fold) by reductive alkylation and acylation,[5] to provide extra internal crosslinks.

Abzymes are also promising, but so far we have only been able to perform small-scale demonstrations with these.[6] The future lies in large scale immortalised immune cell culture, or the use of recombinant (genetically mixed) bacteria or yeast to produce the required abzymes.

Food Processing

Certain enzymes—mainly of fungal (including yeast) origin—are used extensively to degrade particular food components. The sugar industry, particularly in the US, makes extensive use of glucose isomerase for converting the weakly sweet sugar glucose to the sweeter fructose isomer. Chemical equilibrium is reached rapidly in this freely reversible reaction, to form the glucose and fructose (1:1) mixture familiar as invert sugar. It is possible to shift the composition of the product in favour of fructose—this is often referred to as high fructose syrup.[7] Much effort has been devoted to improving glucose isomerase by protein engineering—its stability is increased by replacing Lys253 by arginine, although the mechanism of this is not fully understood. Invert sugar is made by the hydrolytic action of the yeast enzyme invertase—or by using acidic ion exchange resins—on table sugar, sucrose.

In converting starch to glucose, highly viscous concentrates are obtained by solubilising starch with the pH-incompatible enzymes, α-amylase and glucamylase. The inside of the negatively charged enzyme immobilisation supports concentrate protons and therefore have a lower pH than their surroundings; conversely, positively charged supports have a higher internal pH. Manipulating the apparent optimum pH of enzymes by using different supports is commonplace, but this method can give lower yields if using pH-incompatible mixtures of enzymes. Ensuring that these pH optima are compatible is clearly a target for redesigning some industrial enzymes.

In addition, novel delivery systems are being developed for various enzymes that are used in the food industry. For example, cheese-ripening enzymes can be encapsulated in liposomes, which are then added to the cheese milk. These liposomes help stabilise then enzymes, and give a uniform enzyme distribution. The liposomes then gradually break down, releasing the enzymes so they can speed up the ripening of the cheese.

Brewing with Yeast

In the brewing industry it is currently preferable to use whole yeast cells, perhaps immobilised, rather than the enzyme mixtures obtained after yeast disruption, for catalysing the conversion of glucose to ethanol. Soluble enzymes no doubt cluster together in the living cell, retaining any unstable intermediates, and thus immobilised enzymes should ideally be fixed in the same sequence around the supporting particle as in the naturally occurring biochemical pathway in yeast. Such spatial fixing of enzymes requires the manufacture and use of tailor-made supports that exhibit the correct sequence of functional groups. In addition, each enzyme would need to be chemically activated to fit specifically one of these points of attachment. This is possible with enzymes because several amino acids are easily activated. Some of these are known to be more reactive than others, especially if they occur on the surface of the enzyme, and hence we can specifically activate particular amino acids. We could also redesign naturally occurring enzymes to accommodate specific immobilisation points for binding.[8] Nevertheless, brewing with whole yeast cells has many other advantages, including flavour enhancement.

The Complete Enzyme in Use
Many of the most useful enzymes are glycoproteins that contain a variety of oligosaccharides. Currently, only the incomplete protein structures are available through genetic engineering, and enormous effort is now being devoted worldwide to achieving the complete glycosylated structures. Yeast normally makes glycosylated enzyme, but unfortunately its mode of glycosylation is not suitable for producing fully active and stable mammalian enzymes.

The rapidly growing use of hemicellulases and cellulases—obtained cheaply from fungal sources—is introducing further selectivity in hydrolysing the corresponding polysaccharides. In theory, we could use enzyme mixtures—'cellulases'—to convert the massive quantities of sawdust that accumulate in the timber industries to glucose. But in practice this procedure is hampered by an inability to expose the cellulose in the wood ready for attack by cellulases. Lignin peroxidase isolated from white wood rot fungi, for example, only poorly degrades the lignin barrier shielding the cellulose. Lignin peroxidase has some mechanistic resemblance to the family of monooxygenase (mixed function) enzymes known as cytochromes P-450. These could be useful in the pharmaceutical and food industries, especially as particular mammalian forms become commercially available after genetic engineering.[9] The range of cytochromes P-450—over 200 types, with widely different specificity—obviates any immediate requirement for enzyme redesign, except to improve poor thermostability. Enzyme thermostability is a consequence of their rigid conformation, which is in turn dictated by the extent of hydrogen and electrostatic bonding within the protein structure. Thus, we can greatly improve the thermostability of many enzymes by effecting subtle changes in their amino acid sequences. A loss of enzyme activity is associated with the unfolding of protein globular structure; reversal is possible, however, by refolding.

The main large-scale bioconversion step in the pharmaceutical industries is the conversion of penicillin G to 6-aminopenicillanic acid. This reaction is done at pH 8, by using the enzymes penicillin acylase, amidase or deacylase from *E. coli* to catalyse the removal of a phenylacetic acid side chain. New side chains of penicillins can be added chemically or enzymatically by using the same enzyme 'in reverse' at pH 6. Columns of the enzyme covalently immobilised to, for example, DEAE-cellulose are used to prevent any bacterial enzyme from contaminating the new penicillin. The use of positively charged supports decreases the apparent pH optimum, and designer-improved features can achieve efficient use of such an enzyme.[10]

Enzyme Mimics
Cytochromes P-450 have the potential for a variety of future uses as stereo- and regiospecific enzymes in biosynthesis and bioanalysis.[11-13] They are called mixed function oxidases because of their mechanistic ability to split O_2 and insert one atom of oxygen into the substrate (usually to form a hydroxyl group); the other oxygen atom combines with the two protons to form one molecule of water. Cytochromes P-450 can generate oxygen free radicals in the absence of substrate, causing non-specific damage of cell macromolecules including proteins and DNA, leading to cell injury and death. Some substrates of cytochrome P-450 such as carbon tetrachloride and chloroform also generate toxic free radicals.

Enzyme mimcry[4,14-16] requires a detailed knowledge of enzyme mechanisms, which is often unavailable. The protein moiety of an enzyme provides the observed regio and stereospecificity, and identifying the composition and configuration of the active site is often insufficient to appreciate an enzyme mechanism fully. Most mimics fail to approach the rate constant of the corresponding enzymes, except in the simplest model systems. Nevertheless, the design and synthesis of heat and pH-stable mimics for use in industrial processes is an attractive goal for chemists and chemical engineers. The simple cytochrome P-450 mimic is an iron-porphyrin complex,[17] with an appropriate structure to resist oxidative destruction by the organic hydroperoxides and other oxidising agents likely to be used instead of molecular oxygen in the formation of metal-bound oxygen species. A substrate binding site must be attached to this oxygen activating centre. In cytochromes P-450 this contains the fifth ligand to the iron, a thiolate (-S-) moiety of cysteine. The other four ligands to the iron atom are the nitrogen atoms of the surrounding protohaem IX, while the sixth ligand is usually considered to be water. The oxygen binds by displacing the thiolate moiety at the fifth ligand (iron is in the Fe(II) form). So far the mimicry of cytochromes P-450 and other enzymes has been of limited success because of our poor knowledge of the subtle conformation changes adopted by the enzyme in reaching the transition state for a variety of substrates that require regio- and stereospecificity in their biotransformation.

Production of process-improved enzymes will be achieved firstly by protein engineering and protein tailoring techniques[18-20] and each novel biocatalyst will be patented for use in particular bio-process industries, often with pharmaceuticals or foodstuffs.

References

1. A. Wiseman and H. Dalton, *Trends Biotechnol.,* 1987, **5,** 241.
2. A. Wiseman, *J. Chem. Tech. Biotech.,* 1993, **56,** 3.
3. A. Wiseman, *Handbook of enzyme biotechnology,* 3rd edn. London: Ellis Horwood, in press.
4. R. Breslow and R. Xu, *Proc. Natl. Acad. Sci. USA,* 1993, **90,** 1201.
5. V.V. Martinek *et al, Biotechnol. Bioeng.,* 1992, **40,** 650.
6. S.A. Lesley, P.A. Patten and P.G. Schultz, *Proc. Natl. Acad. Sci. USA,* 1993, **90,** 1160.
7. P. Goodenough, *Biologist,* 1993, **40,** 67.
8. A. Wiseman, *Trends Anal. Biochem.,* 1992, **11,** 303.
9. A. Wiseman, *Genetically-engineered proteins and enzymes from yeast: production control.* London: Ellis Horwood, 1991.
10. T. Barenschee, T. Scheper and K. Schuegerck, *J. Biotechnol.,* 1992, **26,** 143.
11. A. Wiseman, *Trends Biotechnol.,* 1993, **11,** 131.
12. T.L. Poulos in *Methods in enzymology,* vol 206, M.B. Waterman and E.F. Johnson (eds), p 11, London: Academic Press, 1991.
13. H. Ohkawata *et al* in *Opportunities in biotransformations,* L.G. Copping *et al* (eds) p 23. Cambridge: Elsevier, 1990.
14. R. Bar, *Trends Biotechnol.,* 1989, **7,** 2.
15. S. Borman, *Chem. Eng. News,* 1992, **70,** 4.
16. J.M. Stewart, K.W. Hahn and W.A. Klis, *World patent app.* 91/10733, 1991.
17. D. Mansuy and P. Battioni in *Bioinorganic catalysis,* J. Reedijk (ed) p. 395. New York: Marcel Dekker, 1992.
18. M. Winkler and A. Wiseman, *Biotechnology and genetic engineering reviews,* vol 10, M.P. Tombs (ed) p 185. Newcastle upon Tyne: Intercept, 1992.
19. A. Wiseman, *Endeavour,* 1992, **16,** 190.
20. R. Breslow in *Host-guest molecular interactions from chemistry to biology,* p 115, Ciba Foundation symposium 158. New York: John Wiley, 1991. ∎

Questions

1. Why has the use of food enzymes in food processing been delayed?

2. What are the positive effects of encapsulating cheese-ripening enzymes in liposomes?

3. Why hasn't the theory of using cellulase enzyme mixtures to convert sawdust to glucose been successful yet?

Answers begin on page 161.

35 *Geothermal vents and hot springs contain microbial lifeforms that can proliferate at temperatures above 100°C. Whereas the optimum temperature of enzymes of most organisms falls in the range of 25–40°C, the enzymes within these microorganisms function efficiently at much higher temperatures. Aside from their biological significance, enzymes play vital roles as catalysts in a number of chemical industries. Because many chemical processes give better product yields at elevated temperatures, these hyperthermophilic enzymes offer great potential for a number of commercially important uses.*

Life in Boiling Water

Robert M. Kelly, John A. Baross, and Michael W.W. Adams

Apart from pandering to our scientific curiosity, studies of microbial lifeforms that can exist at temperatures above 100 °C are revealing a host of commercially important enzymes.

It is interesting and at the same time ironic that one of the new frontiers in the biological sciences focuses on microorganisms that are among the most ancient of life forms on this planet. Microorganisms that proliferate under extremes of temperature, pH, pressure and salinity often come from environments that mimic the earliest conditions of life on earth. Indeed, it has been argued that the first life forms were thermophilic (heat-loving) and that most extant life has evolved to what we now consider ambient conditions. Many of these high temperature organisms have only been isolated relatively recently and efforts to explore geothermal environments for basic scientific as well as technological reasons are on the increase.

The thermal distribution of life in the biosphere, as it is currently understood, is strongly biased to moderate temperatures, between 20 and 40 °C. The inhabitants of the highest temperature climes—microorganisms growing above 90 °C—are almost exclusively from the kingdom known as Archaea. These are prokaryotes (lack a cell nucleus) and have many metabolic similarities to the Bacteria. Their evolutionary placement in the phylogenetic tree proposed by Carl Woese and co-workers at the University of Illinois[3] at Urbana-Champaign in the US is based on changes in nucleotide sequence of a certain type of RNA. Differences in this sequence can be related to evolutionary distance, suggesting that the Archaea are closer to the Eucarya—organisms that possess a nucleus—than the Bacteria. Additional information on the patterns of metabolism and the constituent biomolecules of the Archaea should help define their evolutionary significance. The extreme conditions (high temperature, high salt, low pH) in which they grow distinguishes them from most other microbes. Among the thermophilic Archaea are organisms growing at up to 113 °C; those that can grow above 90 °C will be referred to as hyperthermophiles. To date, all of these species have been isolated from geothermally heated environments.

Geothermal Habitats

With the discovery of deep-sea hot springs—the most striking examples being sulphidic chimneys (hydrothermal vents) that spew superheated fluids into the cold ocean—it is clear that the Earth's geothermal activity is widespread. Although there are marked differences between marine and terrestrial hot springs from which hyperthermophiles are isolated, the geophysical processes driving all volcanic systems are fundamentally the same.

The main differences between terrestrial and submarine hot springs are the higher hydrostatic pressures and temperatures in deep-sea hydrothermal systems, and the diversity of geothermal hot springs in terrestrial environments. The concentrations of some important biological chemicals are also much higher in deep-sea hydrothermal vents. The concentration of H_2S, for example, can range from 3.5 to 8.4 mM in hot hydrothermal fluids, whereas H_2S concentrations rarely reach 1 mM in terrestrial hot springs. Similarly, methane concentrations as high as 7 mM have been reported at some deep-sea hydrothermal vent sites in the Pacific Ocean at the Guaymas Basin in the Gulf of California and at the Endeavour Segment of the Juan de Fuca Ridge off the coast of Washington and Oregon. The methane levels in fumaroles (volcanic vents) and terrestrial hot springs are usually a few parts per million, but can be considerably higher at solfatara seep sites—hot sulphur-containing water that oozes from the ground—if there is an active methanogenic microbial community. In contrast, H_2 concentrations can be as high as 5 per cent of the total gases in fumaroles, but generally less than 10 nM in terrestrial hot springs. Low levels of H_2 generally correlate with high levels of methane in both terrestrial and submarine hot springs.

Other biologically important species such as nitrogen, phosphorus, organic compounds and transition metals (especially iron, manganese, molybdenum and tungsten), are found in these environments. While metal concentrations are generally high in all terrestrial and submarine hot springs, the concentration of nitrogen compounds varies considerably, depending on the amount of organic material. There are seep sites in Grand Teton National Park in the US with nitrogen contents from 69 to 90 per cent of the total gases. Little is know, however, about the sources, levels and types of phosphorus and organic compounds in most submarine hydrothermal systems. Guaymas Basin in the Gulf of California is one of the few submarine hydrothermal vents having high levels of organic material. This is attributed to a significant amount of phyto-plankton-derived material sinking to the depths, forming a thick organic sediment layer. The organic material is pyrolysed, resulting in high concentrations of hydrocarbons and other petroleum-like compounds. Even though CO_2 is the predominant carbon gas at Yellowstone Park hot springs, hydrocarbons have been detected at about 1 per cent of the total carbon content, occasionally as highly paraffinic crude oil.

Researchers have detected or isolated microorganisms from most terrestrial and submarine hot spring environments where the temperatures are within the range for growth or survival.

The most studied geothermal environments at submarine hydrothermal vent sites are the so-called 'black smoker' and 'flange' formations, together with the associated hot fluids and sediments. These structures form when metal sulphides precipitate from the hot thermal fluids on contacting the cold ocean water. We have isolated hyperthermophilic microorganisms from these smokers and their environs as well as from sulphide chimneys found at some vent sites. A characteristic of these large sulphide structures is the presence of outcrops, called flanges, under which there is a pool of hot hydrothermal fluid. A small portion of the hot fluid diffuses upwards through the flange, creating an internal temperature gradient. The remaining fluid spills over the edge of the flange, creating small smokers. The surface of the flange still contains some intact vent animals; the cavity in the lower portion is a pool of 350 °C fluids. Large numbers of hyperthermophilic Archaea—sulphur dependent heterotrophs and methanogens—have been detected and isolated from these flanges.

A Cultural Revolution

Most studies are restricted to overcoming the problems of growing hyperthermophiles in the laboratory, and research into the applied and scientific aspects is limited.[4] Although the high hydrostatic pressures of deep-sea vents can affect physiological and biochemical properties,[5] all of the hyperthermophiles studied to date can be cultured at ambient pressures. Most reduce sulphur to sulphide during growth, creating highly corrosive and toxic mixtures, not to mention a potentially unpleasant working environment. However, hyperthermophilic enzymes are mainly purified from organisms that do not require sulphur for growth; for example *Pyrococcus furiosus*,[6] *Thermococcus litoralis*[7] and *Termotoga maritima*.[1] Thus we can use large-scale stainless steel fermentation systems. To produce enough enzyme for evaluating the biotechnological potential, many researchers are trying to clone the genes for these enzymes and insert them in more easily cultivated microorganisms. To date, few active hyper-thermophilic enzymes have been successfully produced in recombinant microorganisms. We need to culture hyperthermophiles in sufficient quantities to allow us to measure enzyme activities and/or to isolate the DNA.

There are many reasons for choosing to study a particular hyperthermophilic enzyme. In some studies, researchers are focusing on trying to understand microbial physiology at high temperatures. For example, they have characterized the main glycolytic or energy generating pathways and purified several of the enzymes. Other researchers are interested in finding out about high thermostability, and are studying enzymes that are biocatalytically similar to those with activities at much lower temperatures. Recently, chemists have been pursuing enzymes with potentially important commercial uses such as those isolated from *Pyrococcus furiosus* and *Thermotoga maritima*. The high thermostability of these enzymes is reflected in their long half-lives at high temperatures. More successful cloning of the genes for hyperthermophilic enzymes in better host organisms would help us to evaluate their potential. Nevertheless, several interesting developments hold promise for biocatalysis at elevated temperatures and for understanding extreme thermostability.[16,17]

Thermostable Polymerases

Currently, the most important hyperthermophilic enzymes are the thermostable DNA polymerases, which are used in Cary Mullis' recent Nobel prize-winning technique for copying specific stretches of (or selectively amplifying) DNA—the polymerase chain reaction (PCR). This involves heating samples of DNA containing the sequences to be amplified, causing the double helix to split into single strands. We then add short base sequences (primers), which attach to the complementary bases flanking the region that is about to be amplified. Hence the DNA polymerase is restricted to copying only the specific base sequence that is of interest, resulting in two identical double helices, ready to repeat the PCR cycle. Because of its thermostability, there is no need to add more polymerase after each cycle—which can reach temperatures[18] close to 100 °C. Several DNA polymerases from hyperthermophiles are now commercially available. Unlike the commonly used Taq polymerase from Thermus aquaticus,[19] these new enzymes have high levels of activity; by 'proofreading' the sequence of nucleotide bases added to the growing strand of copied DNA they can produce higher fidelity copies of the template. For example, the DNA polymerase from *Pyrococcus furiosus,* has a half-life of over 20h at 95 °C and shows 10-fold improvement in fidelity over Taq polymerase after a 105-fold amplification in the cases studied.

Thermostable Isomerases

One potential application of high temperature biocatalysis is to isomerase glucose syrups to a temperature-dependent equilibrium mixture of glucose and fructose. This conversion is essential in the production of high fructose corn syrup. It has been shown that the glucose isomerase from the hyperthermophile *T. maritima* is optimally active at about 100 °C, which is significantly higher than the current enzymes[13] used in processes run at 60 °C. This increased isomerization temperature would allow us to produce syrups with higher equilibrium fructose concentrations, giving a significant process advantage. Several attempts are under way to characterize glucose isomerases from several *Thermotoga* species and include determining the gene sequence and expressing the enzyme in a more easily cultivated host microorganism. Because they are similar to well studied glucose isomerases that are active at lower temperatures these enzymes may also be important models for investigating hyperthermostability.

Model Systems

It was originally believed that gene sequence and structural information on proteins from hyperthermophiles would clearly illustrate strategies for thermal stabilization. This has not been the case; the limited information available on these proteins seems to have ruled out any universal explanation of protein thermostability until suitable crystal structures become available.

Several groups are currently attempting to determine crystal structures for a variety of hyperthermophilic proteins, although only one such protein has been examined with the benefit of a structure, a rubredoxin from *P. furiosus*.[20] Unlike other rubredoxins, the thermostability of the one from *P. furiosus* seems to be related to an ordering at the N-terminus of the protein that allows it to be a part of the hydrogen bonding network of an internal β-sheet structure. One explanation for its stability is that this strong hydrogen bonded network prevents the protein from unwinding when exposed to thermal stress.

Future Work

It might be argued that the most profound impact of research on hyperthermophilic organisms will be in revealing some of the factors contributing to biomolecule stabilization at extreme temperatures. However, we should derive many other potential benefits from these novel microorganisms. The current interest in the origins of life has led many researchers to examine the hyperthermophiles because of their ability to grow under the extreme conditions thought to prevail when life began. These organisms are also likely to be a useful source of genes coding for a variety of commercially important enzymes. Whole cell biotransformations at elevated temperatures with pure and mixed cultures of hyperthermophiles may be the basis for successful bioremediation processes. For these and other reasons, the scientific and technological communities should continue to explore how these ancient life forms could provide solutions to problems that need to be addressed now and in the future.

Acknowledgments

The authors wish to thank the US National Science Foundation, the Department of Energy, the Office of Naval Research, the National Institutes of Health, and the North Carolina Biotechnology Center for supporting this research.

References

1. R. Huber *et al, Arch. Microbiol.,* 1986, **144,** 324.
2. *Idem,* 1992, **154,** 340.
3. C.R. Woese, O. Kandler and M. L. Wheelis, *Proc. Natl. Acad. Sci. USA,* 1990, **87,** 4576.
4. R.M. Kelly *et al* in *Biocatalysis at extreme temperatures* (M.W.W. Adams and R.M. Kelly, eds), ACS symp. ser. no. 498, p 23, 1992.
5. C.M. Nelson, M.R. Schuppenhauer and D.S. Clark, *Appl. Environ.*

Microbiol., 1991, **57,** 3576.
6. G. Fiala and K.O. Stetter, *Arch. Microbiol.,* **1986,** 145, 56.
7. A. Neuner *et al, Ibid,* 1990, **153,** 205.
8. I.I. Blumentals, A.S. Robinson and R.M. Kelly, *Appl. Environ. Microbiol.,* 1990, **56,** 1992.
9. S.H. Brown and R.M. Kelley, *Appl. Environ. Microbiol.,* 1993, **59,** 2614.
10. E.J. Mathur *et al, Nucleic Acids Res.,* 1991, **19,** 6952.
11. F.O. Bryant and M.W.W. Adams, *J. Biol. Chem.,* 1989, **264,** 5070.
12. H.R. Badr, K.E. Sims and M.W.W. Adams, *Syst. Appl. Microbiol.,* in press.
13. S.H. Brown, C. Sjoholm and R.M. Kelly, *Biotechnol. Bioeng.,* 1993, **41,** 878.
14. H. Simpson, U.R. Haufler and R.M. Daniel, *Biochem. J.,* 1991, **277,** 413.
15. J. Gabelsberger, W. Liebl and K.H. Schleifer, *FEMS Microbiology Lett.,* 1993, **109,** 131.
16. M.W.W. Adams, *Annu. Rev. Microbiol.,* 1993, **47,** 627.
17. R.M. Kelly and S.H. Brown, *Curr. Opin. Biotechnol.,* 1993, **4,** 188.
18. R.K. Saiki *et al, Science,* 1985, **230,** 1350.
19. K.B. Mullis and F.A. Faloona, *Methods Enzymol.,* 1987, **155,** 335.
20. J.E. Wampler *et al, Protein Sci.,* 1993, **2,** 640. ∎

Questions

1. What distinguishes organisms in the kingdom Archaea from other microbes?

2. What problems have been encountered when trying to grow hyperthermophiles in the lab?

3. Why is the thermostable DNA polymerase preferable for use in the DNA polymerase chain reaction (PCR) technique?

Answers begin on page 161.

36 *Human DNA contains approximately 100,000 different genes that determine the genetic characteristics of every individual. Researchers around the world are working to identify each of the genes, a feat that will provide a key to understanding the nearly 4,000 known genetic disorders and countless diseases that may originate in part with genetic malfunctions. Two scientists have revolutionized the process for gene sequencing, and with their discovery an explosion of gene identification is taking place. A number of important diagnostic products may soon emerge from their work.*

The Gene Kings

John Carey, with Joan O'C. Hamilton, Julia Flynn, and Geoffrey Smith

Two scientists are changing how DNA is mined—and drugs are discovered.

Freud was wrong: The gene, not anatomy, is the closest biology comes to destiny. The 60,000 to 80,000 genes in each of our cells are the blueprint—the operating system, if you will—of humanity. Inscribed in their double helixes of DNA are tales of life and death, sickness and health. By instructing cells to make proteins, they determine whether our eyes are blue or brown, whether or not we can dunk a basketball, and whether we should worry about developing heart disease or breast cancer.

What's turning pharmaceutical executives into gene aficionados is the prospect of vast commercial treasure in those spiraling strands. A single gene, if it makes a protein that works as a drug, such as Amgen Inc.'s anemia-fighting Epogen, could mean a product worth more than half a billion dollars a year. And that's just the beginning. Tests to spot the faulty pieces of DNA that cause such diseases as cystic fibrosis or colon cancer—a market worth $376 million a year—should balloon into a multibillion-dollar business early next century. Greater knowledge of human genes promises "smarter" drugs and the ability to nip disorders in the bud. Genes "are the raw material for the next wave of therapeutic discoveries," says Lawrence Livermore National Laboratory's Gregory G. Lennon.

Five years ago, science's DNA sleuths—most of them academic researchers—had nabbed less than 5% of all genes. And it appeared that reading the rest would take many years. One man changed all that: J. Craig Venter, an ex-surf bum and obscure National Institutes of Health scientist, perfected a method to rapidly find and sequence genes.

Venter still seeks respect in the snobbish world of science. But his innovations are transforming the pharmaceutical industry. Even skeptics such as Dennis Henner at Genentech Inc. admit Venter's work is "making companies reevaluate all their technology and drug discovery methods." The impact of this genetic information "is going to be enormous," adds Upjohn Co. Distinguished Scientist Jerry Slightom. "It may be the mainstay of drug companies in the future." It is also turning Venter and fellow scientist William A. Haseltine into potential Gene Kings.

The two men have formed one of the oddest alliances in biotech: Venter, 48, heads the nonprofit The Institute for Genomic Research (TIGR) in Gaithersburg, MD. Haseltine, 50, is CEO of Human Genome Sciences Inc. (HGS) in nearby Rockville, MD., which has rights to commercialize TIGR's findings. Together, they have deciphered DNA sequences representing parts of a staggering 85% to 90% of all human genes. The functions of more than half of these remain unknown. Still, they say, their databases contain leads for scores of new drugs. To Haseltine, their discoveries rival that of Balboa, who crested a mountain range in Panama to see—and claim for Spain—a whole new world. "This isn't like oil—there's not more than one gene pool," Haseltine exults. "All people who come later can do is repeat what we've done."

**Reprinted from the May 8, 1995, issue of *Business Week* by special permission.
Copyright 1995 by The McGraw-Hill Companies.**

Critics see that as classic hype from Haseltine, who is known as much for his hubris as for his scientific brilliance. But giant drugmakers and startups alike have anted up millions of dollars in a frenzied race to acquire and mine this genetic treasure trove. Smith-Kline Beecham was the first believer. It committed $125 million to HGS in May, 1993, for a 7% equity stake and first dibs on promising genes. Now, everyone wants to be king. Hoffmann-La Roche Inc. has invested $70 million in Millennium, a Cambridge (Mass.) startup. Upjohn and Pfizer Inc. have paid millions to look into the data banks of HGS's main rival, InCyte Pharmaceuticals in Palo Alto, CA. And last September, Merck & Co. funded a major gene-sequencing operation at Washington University.

The Real Genetic Jackpot
Experts see Merck's approach as a direct attempt to undermine the Gene Kings. The drug Goliath will make all the gene-sequencing data public. To some, it's the biotech equivalent of a computer company giving away its new operating system—and making money on the applications. Merck executives figure that if everyone has the same information, the company's vaunted research and development department can win most of the races to market. "Making drugs from genes is like going from a dictionary to the works of Shakespeare," explains Merck's Alan R. Williamson, vice-president for research strategy worldwide.

The great debate, however, goes beyond HGS vs. Merck, or proprietary vs. "open" gene strategies. The real genetic jackpot may have nothing to do with the sheer number of genes identified and cataloged. That's why, rather than blindly sequencing tens of thousands of unknown pieces of DNA, outfits such as Millennium and Sequana Therapeutics are hunting for genes that cause diseases such as diabetes and obesity.

Even this more targeted approach, critics say, doesn't guarantee success. After all, the world is awash in DNA data, but brilliant new therapies are rare. "Identifying genes is only the beginning of a long, painful, and expensive process of drug development," explains Michael Steinmetz, vice-president for clinical R&D at Hoffman-LaRoche. In the end, says skeptic Stephen G. Pagliuca, managing director of Bain Capital Inc., a Boston consulting firm: "Investing in genomics is like going to Las Vegas."

Venter and Haseltine are unfazed. They've deployed powerful super computers to pin down functions of thousands of unknown genes and, with SmithKline, pull out those that could lead to products. HGS is also madly filing patents on everything from gene fragments that help diagnose cancer to proteins with possible therapeutic benefits. Even where HGS and SmithKline opt not to develop a product, the patents may allow them to reap royalties on others' drugs.

A Golden Record
This aggressive patent stance has incensed rivals. "They want to be the gatekeeper of the genome," fumes one pharmaceutical executive. "They think they have everyone else over a barrel." It may also have incited Merck's radical actions. But Venter belittles the threat from the pharmaceutical giant. "They're just validating my approach," he says, "except they're a couple of years late." Even when the Merck project ends in 1996, says HGS's drug-development chief, Michael J. Antonaccio, HGS and TIGR will still have a far thicker "dictionary" of genes. Merck "may use its dictionary to come up with some very good poetry," he says, "but our potential is so much greater."

If HGS lives up to that potential, much of the credit will got to Venter. A champion backstroker in high school, "I was anti-intellectual to an extreme," he recalls. He shunned college to surf Newport Beach, while working as a night clerk at Sears, Roebuck & Co. As the Vietnam War draft loomed, Venter enlisted in the Navy—with the understanding that he would be on the swim team.

When Venter was in boot camp, President Johnson escalated the war and shut down military sports teams. Fortunately, Venter recounts, he scored highest out of 30,000 on his military intelligence test and got his choice of training. He picked hospital corpsman because it wouldn't extend his three-year enlistment. That left him patching up wounds in a Navy hospital in Da Nang and in a Vietnamese village. "It was a lifetime of education packed into one year," he says.

When he left the Navy, a now academically driven Venter raced through his undergraduate degree and his PhD in biochemistry in six years. He quickly snared a faculty position at the University of California at San Diego and, in 1984, was recruited by the NIH.

If Venter's background seems unusual for a scientist, Haseltine is pure pedigree. A Navy physicist's son, he grew up steeped in science. His life has been a golden record of accomplishment: high honors at the University of California at Berkeley, training with Nobel laureates James D. Watson and Walter Gilbert, a prestigious post at Harvard University, and discovery of several HIV genes.

With that, Haseltine leaped into the limelight. A leading spokesman on AIDS, he lobbied congress for more funding, warning the disease would spread to heterosexuals. Critics called him a publicity hound. "I believe it's your responsibility to speak out if you see a major health problem, even if that poses a risk to your own career," he replies. He also ruthlessly exposed flaws in others' work while extolling his own. "His personality is a real liability," says friend and Harvard colleague Max Essex. Were his manner more diplomatic, Essex adds, "he could be ruling the world." Haseltine shrugs off the criticism. "Some people interpret enthusiasm as arrogance," he retorts.

But Haseltine also earned respect for his research and his ability to nurture talent in others. "He was a powerful mentor," says former student Alan D'Andrea, an associate professor at Harvard Medical School. "I was always glad I was on the same side of the table."

While Haseltine thrived in high-profile science, it was Venter who took an obscure idea and transformed it into a critical breakthrough. In 1986, he spotted a paper by geneticist Leroy E. Hood that suggested a method for DNA-sequencing with robots instead of manually. Having spent months doing it the old-fashioned, tedious way, "I was one of the few people who got excited," Venter recalls. He scrounged $110,000 in funding to get one of the first sequencing machines from Applied Biosystems Inc., developed from Hood's ideas. And Venter was the first scientist to get it to work, he says.

Recast as a Villian
It was a time of growing excitement and controversy in biology as scientists were proposing the mammoth 15-year, $3 billion Human Genome Project. The idea was to first map every part of the 23 pairs of human chromosomes, using known genes and other types of markers to place signposts along the chromosomes. Next, scientists planned to sequence all of the DNA in between the signposts. Venter's idea—using the new machine to plunge into the second part of the genome project by reading big chunks of the X chromosome— won support from Watson, the co-discoverer of the structure of DNA, who was heading the NIH's genome project. But review committees repeatedly denied it funding because the technology was deemed too new.

Then Venter had a better idea. Although each human cell harbors some 3 billion individual molecules of DNA, strung together like rungs of an immense ladder, only 3% of them are found in actual genes. These instruct cells to make specific proteins, which in turn, control how the body develops and functions. The rest of the genome is so-called junk DNA, with largely unknown functions.

If he just read the code spelled out by every piece of DNA, Venter figured, the presence of all that junk would make it nearly impossible to find—or understand—the genes themselves. But, as he says, "cells are smarter than scientists." In order to create necessary proteins, cells ignore the junk and copy the DNA of important genes into a related molecule, messenger-RNA (mRNA), which tell the cells' protein what to make. "My insight—or my perception of the obvious—was recognizing that biology works pretty damn well," he says.

To pinpoint the genes, Venter planned to scoop up the easily spotted but fragile mRNA, then make sturdier DNA copies—cDNA—to feed into his sequencing machines. To speed things even more, he read just part of the DNA, since each interesting fragment could later be used to fish out the whole gene. The approach cut the cost of sequencing an unknown gene from an estimated $50,000 using older methods to roughly $20.

It wasn't a new idea, and the gene-hunting elite didn't buy it. They argued that nature was essentially uncooperative: Only a tiny fraction of the genes in a given group of cells are turned on. For example, muscle cells might be making only a couple of key muscle proteins, leaving everything else switched off. They figured the method would snare mRNA from 8% of the genes, at most—many of them uninteresting.

They were wrong. "Venter lucked out," says University of Wisconsin gene-sequencer Frederick R. Blattner. "Nature was not throwing us the terrible curve ball that people feared." Other scientists were spending years to collar one gene; Venter was able to sequence bits of 100 human brain genes in months.

Then, all hell broke loose. Figuring the sequences had value that should be protected, the NIH filed patent applications on 315 gene fragments in June, 1991. It later withdrew all the applications, but not before Venter had been recast as a scientific villain. Part of the flood of criticism was aimed at the idea of patenting gene fragments, but much was a direct assault on Venter and his method. Watson charged at a July, 1991, Senate hearing that Venter's operation "could be run by monkeys." Other luminaries sniffed that Venter hadn't found anything major, such as the Huntington's disease gene, and that what he was doing was uninspired, unsporting, and a threat to the genome project. "I had three strikes against me," Venter recalls. "I had a radical idea, it worked, and I was an outsider."

Even now that cDNA has become mainstream, many scientists belittle Venter's contribution. "He gets credit for inventing this whole technology—that's blatant nonsense," says Hood. "He has never invented anything. The only thing he deserves credit for is scaling up the process." Retorts Venter: "What really pisses people off is once they see it works, they think: 'I could have done that.'"

But if scientists failed to applaud, the business world took notice. At a time when only a few human genes were known, "Craig offered an approach that might enable you to quickly discover most of the rest," says venture capitalist Wallace H. Steinberg, chairman of HealthCare Investment Corp., which founded TIGR and HGS. "It was not an opportunity that would ever come along again."

By May, 1992, Venter had two $70 million offers: one from Steinberg and one from Amgen. Steinberg was ready to commit only $20 million. But Venter wanted $70 million (later upped to $85 million) for a nonprofit outfit that would let him do the academic work he hoped would boost his standing in the scientific community. Given the tremendous value of the approach, Steinberg says, he had no choice but to ante up. At the same time, he founded HGS to commercialize TIGR's gene sequences and began looking for a CEO to run it.

Enter Haseltine. Even his numerous critics readily admit he is an exceptional scientist. "Bill has a smell for what's important in molecular biology," says NIH virologist Robert C. Gallo, co-discoverer of HIV. In fact, venture capitalist Steinberg had launched several Boston-area companies based on the Harvard virologist's ideas. But he couldn't induce Haseltine to leave the ivory tower until HGS came along. Haseltine took a leave from Harvard to move—with his socialite wife, Giorgio perfume creator Gale Hayman—to Washington.

Meanwhile, Steinberg began approaching drug companies, trying to sell Venter's gene data. Rhone-Poulenc Rorer was one company that declined. "We couldn't handle all that data," explains one insider. SmithKline R&D chief George Poste, on the other hand, recognized "a technology that was absolutely fundamental for our future competitiveness." Despite widespread skepticism in the drug industry, he persuaded his bosses to sign on. "It was a big gamble," says Stanford University biochemist Paul Berg.

On-Switches
The colossal question now is: how to go from the hundreds of thousands of gene fragments with unknown functions to products. As always, skeptics abound. The DNA code of those fragments offers "no insights into anything," says Hood.

Some scientists, though, see tantalizing clues in the sequences of the cDNA fragments. They have learned to discern key features in snippets of genetic code. For example, one pattern, or motif, of individual DNA molecules usually results in a protein attached to a cell membrane, such as a receptor for outside signals. With an unknown gene, says Berg, "it's possible to make a good guess at the function of the protein it encodes." The guess may sometimes be wrong, so HGS fully sequences scores of its genes, testing the proteins they make.

In addition, academic researchers have been making enormous strides in probing the DNA of simple critters such as yeast and nematodes, whose genes usually have human counterparts. When the machines at HGS spit out a new sequence, scientists comb their databases for telltale motifs or similarities to known genes. SmithKline scientists have used this approach to discover what Poste describes as "a totally novel enzyme" that dissolves bone. He hopes a chemical that inhibits the enzyme may offer a treatment for osteoporosis.

The strategy works in reverse, too. In December, 1993, Johns Hopkins University oncologists Bert Vogelstein and Kenneth W. Kinzler were racing to catch up with rivals at Harvard and the University of Vermont in a hunt for the gene that, when flawed, caused inherited colon cancer. They suspected that the gene normally fixes errors made when cells copy DNA during cell division and had in hand such a gene from yeast. So, in what many see as a vindication of the cDNA approach, they agreed to cede product rights in return for access to HGS's database. In minutes, they found the human version. "It's the way things are going to be done in the future." says Kinzler. That's especially true for the pharmaceutical industry's lifeblood: drug discovery. In the past, drugmakers tested thousands of chemicals to find one that eased pain or lowered blood pressure. More recently, they have begun to grasp the underlying biology, finding enzymes and biochemical pathways that can be blocked or activated by drugs. But the body typically has substitute pathways, so drugs often don't function as well as expected. With a database of cDNA sequences, "you can start with the full set [of genes and pathways] and choose the best one," says HGS Senior Vice-President for R&D Craig A. Rosen.

cDNA could help solve another major problem: a dearth of novel drug targets. Experts believe a database of all human genes must be laden with clues to previously unknown biochemical pathways that could be manipulated to treat or prevent disease. The trick is finding them. One clever method is comparing cells from different organs, or from normal vs. diseased tissue. When scientists at SmithKline studied stroke in animals, for example, they found more than 20 genes that switched on at the onset of the illness and more than 30 that switched off. Poste expects the key to treating or preventing stroke lies hidden in those 50-plus genes.

Similarly, scientists at HGS and TIGR have found that a whole new batch of genes turn on when normal prostate cells turn cancerous. HGS and SmithKline are now using the genes to make diagnostic tests for spotting signs of malignancy. Moreover, some of the genes could offer new strategies for halting cancer in its tracks. The finding, says Samuel Broder, recently departed director of the National Cancer Institute, "is fabulously interesting."

Venter is also pushing the frontiers of basic science. In late May, he will announce the DNA code for the first two microbes ever fully sequenced, offering crucial insights into the biology of these bacteria. It will also provide a better understanding of human biology and disease, since nature reused most bacterial genes in complex mammals, including humans. The NIH had declined to fund this effort, claiming his cDNA method was no match for the task. But once agin, Venter confounded popular wisdom. "No one else thought it could be done," says Nobel laureate and molecular geneticist Hamilton O. Smith of Johns Hopkins.

The Perils of Publishing

Few now doubt that the cDNA approach is transforming science. But who will reap the commercial benefits? Haseltine is determined to stay in the lead—even if it requires a change in strategy. Originally, HGS planned to rely primarily on DNA sequences produced by Venter at TIGR. But by 1993, Venter was turning more to basic science—and HGS faced growing competition from rivals such as InCyte, which were also using the cDNA approach to sequence thousands of genes.

So Haseltine changed HGS's course to downplay TIGR's role. He duplicated Venter's sequencing and computer systems on a grander scale and began churning out 750,000 pieces of DNA code a day, while racing to find and patent the most important new genes. He wants HGS to develop some drugs on its own, help SmithKline make others, and license the rest.

The strategy is risky. For one thing, it has helped create a schism between the two Gene Kings. Venter's unusual deal with HGS gave him the right to publish his findings quickly. In return, HGS got commercial rights to all the genes he discovered. At a cost to his academic standing, Venter initially bowed to pressure from HGS to delay publishing and to focus only on human genes. But now, he's making his data public, with major papers coming out soon. The result is conflict. "Craig feels that HGS basically repeated what he did and stole the glory," says one company insider. "HGS sees Craig's desire to publish as undermining their commercial position. Now, they're handing over $8.5 million a year and not getting much in return."

Internal clashes, however, are the least of Haseltine's worries. HGS's plan puts the bold startup against a host of competitors, especially Merck. With sequences to more that 4,000 gene fragments now flowing each week from the Merck-funded operation at Washington University into public databases, "Merck has really broken the monopoly," says Baylor University gene-sequencer Richard A. Gibbs. What's more, other genome companies, such as Myriad Genetics, Sequana, Mercator, and Millennium, are trying a different tack. They see no point in HGS's approach of blindly sequencing thousands of unknown genes. Instead, they are studying families in which the incidence of diseases such as diabetes or cancer is unusually high. That way, they can snare the underlying genetic mechanisms, then develop drugs for those diseases. HGS may win the race to nail many genes, argues Millennium CEO Mark J. Levin, but his own, more focused strategy will be a faster way of getting drugs to market.

And so, the race is on: the Gene Kings and their partners against the rest of a drug industry, which is straining to catch up. A few skeptics notwithstanding, "nearly everyone is into genomics," says Bear, Stearns & Co. analyst David T. Molowa. "Everyone is a believer." The first wave of products—new diagnostics—may be on the market in a year or two, with potentially important drugs emerging by the end of the decade. By then, it may be clear which strategy will carry the day—and whether Haseltine and Venter have managed to stay on their thrones. Whatever the outcome, their position as pioneers in a medical revolution is already secure. ∎

Questions

1. Why did Ventner make copies of mRNA into cDNA instead of just working with mRNA itself?

2. Why does Merck & Co. want to make their results of the genome public?

3. What percentage of human genes have been deciphered by the research companies headed by Ventner and Haseltine?

Answers begin on page 161.

37 *Remarkable technology is available that allows segments of DNA (genes) from one organism to be introduced into the genetic material of another organism. The application of this biotechnology in agriculture holds the promise for a future of exciting advances in improving crops and designing biopesticides. To many, genetic engineering is an unprecedented opportunity to understand and manipulate the secrets of life for the benefit of society. To others, however, this new technology represents a horrific threat to our environment and is the ultimate scientific recklessness.*

Environmental Risks in Agricultural Biotechnology

Peter Kareiva and John Stark

Biotechnology has given rise to many fears within the public domain. Basic ecological experiments should distinguish the real dangers from the imaginary.

Technology, in the form of fertilizers, pesticides, and new crop varieties, has increased agricultural yields several fold over the past hundred years. Indeed, in many parts of the world there is a surplus of food produced; and were it not for the so-called 'green revolution', the Earth's population might have exceeded food limits decades ago (which was the 'doomsday' prediction of some ecologists). However, now there is an even more extraordinary revolution underway in agriculture—the so-called 'biotechnology revolution'.

Genetic engineering has provided unprecedented opportunities for improving crops, designing biopesticides, manipulating microbial communities, and even 'engineering' fish and farm animals. To many this revolution represents the most exciting intellectual endeavor in the history of science; in a sense the secrets of life are being uncovered and manipulated through molecular biology. But to others, genetic engineering represents the ultimate in scientific arrogance and misguidedness—a threat to our environment and a technology with ghastly social implications.[1]

In this article, we avoid the hyperbole and focus on field-tested advances in biotechnology as they pertain to agriculture. Our goal is to sketch the environmental risks associated with those advances, and how ecologists around the world are attempting to assess those environmental risks. However, before turning to specific crops or biopesticides, it is worth trying to understand why biotechnology causes such fear among large sectors of the public, and to examine the scientific basis of those fears in general.

The Legacy of Pesticides and the Frankenstein Myth

Molecular biologists often express shock that environmentalists are so angry with them for working on genetically engineered crops—after all, aren't they are just trying to improve world food production and reduce the use of chemical pesticides? Two forces drive the public's fear of biotechnology: its experience with pesticides and the powerful imagery of Frankenstein and the perils of 'playing God'. For the first, one only needs to remember the zealous promises made on behalf of DDT, now a proven hazardous chemical—advertisements were run in magazines throughout the world showing a cartoon of vegetables, animals and a housewife dancing together and singing, 'DDT is good for me'. Rachel Carson's now famous book, 'The silent spring', was rebuffed by so-called experts as everything from amateurish, to pure folly. We all know how that story turned out, and so does the public: industry fiercely denied the risks of DDT until the evidence was so overwhelming it had to relent.

The real source for the public's distrust of biotechnology, however, concerns the notion of 'playing God' in some horrifically unnatural way through genetic engineering. Anti-biotechnology essays are often illustrated with pictures of bizarre animal-plant monsters, such as the body of a chicken and the head of a

Reprinted with permission from *Chemistry & Industry*, January 17, 1994.

fish (representing in some way the insertion of fish genes into a chicken). Prominent environmental newsletters carry articles with titles such as, 'Are you ready for Frankenfoods'.[2] While these fears are not something that can be examined in a formal risk analysis, they mean that scientists must be extremely thorough when scrutinising the ecological damage that could arise from the release of genetically engineered organisms.

Nearing the Market Place
Dozens of different genetically engineered crops have been field-tested successfully.[3] The traits inserted into these crops include genes designed to protect crops against pests, genes that enhance the resistance of crops, to particular physical stresses such as freezing or saline soils, and genes designed to alter the character or quality of foods. Ecologists are concerned with three major risks in promoting widespread usage of these crops: (i) recombinant genes may escape via pollen and enter the DNA of uncultivated plants through hybridization, thereby creating new weeds or exaggerating existing weed problems; (ii) genetically modified cultivars may themselves escape from cultivation and invade natural communities; and (iii) the use of certain transgenic crops may undermine ecosystem sustainability through their direct effects on microbial communities and soils, or because they encourage agronomic practices that are unsustainable.[4] While in principle outlining these risks is straightforward, actually evaluating them for specific crops has not proven easy.

To appreciate the subtleties of ecological risk analysis, it is worth considering what would seem at first glance to represent two different extremes in 'environmental friendliness': transgenic insect-resistant crops and transgenic herbicide-resistant crops. Insect resistant crops are typically plants engineered to contain genes for *Bacillus thuringiensis* (Bt) endotoxins. Since Bt endotoxins protect plants against pests selectively, their widespread usage could drastically reduce reliance on chemical insecticides. Moreover, Bt endotoxins are not toxic to mammals. The reason for this is that Bt endotoxins are produced as insoluble crystals that dissolve in the alkaline guts of insects, but are harmless (undissolved) in the acidic guts of mammals.[5] Indeed, Bt crystalline spores are one of the most widely accepted and effective 'natural pesticides'; they are a recommended treatment for organically grown vegetables under attack from caterpillars. What could possibly be worrisome about building such a 'green technology' into plants?

First, although direct consumption of undissolved Bt toxins poses no threat to mammals, it is less clear that mammals and general insect predators would be immune to any risks should they consume Bt-poisoned caterpillars or beetles; in particular, the poisoned insects will have already dissolved the Bt toxins and could act as vehicles for transfer of the toxin through food chains.[6] In addition, should Bt crops decompose in alkaline soils, the crystalline inclusions might dissolve and thereby be released for activity against a wide range of soil fauna.[6] If levels of expression for Bt endotoxin production were sufficiently high in plant tissue, and if the biomass of decaying vegetation were large, inputs of these endotoxins into soils through transgenic plants could exceed anything seen from aerial application of Bt spores under current 'organic' practices.

It is also easy to imagine a Bt-protected plant becoming invasive because the absence of herbivores increases the plant's reproductive rate and vigor so much that the plant runs wild. Indeed, the primary reason given for pest status in many introduced weeds is that they have escaped their natural enemies by moving to new continents—genetic engineering could accomplish the same escape without having to move the plant to a new location.

In contrast to Bt plants, crops engineered to be resistant to herbicides have incurred the wrath of environmentalists from the very beginning.[7] After all, when a chemical company that produces a particular herbicide also develops a transgenic plant resistant to that herbicide (for example, glyphosate and glyphosate resistant crops developed by Monsanto),[7,9] isn't the obvious result further environmental degradation by increased herbicide usage? Even this question does not turn out to be simple. For example, glyphosate is considered one of the safest herbicides on the market today; thus glyphosate-tolerant crops could lead to farmers replacing more hazardous chemicals with less hazardous chemicals.[8] Even resistance to more toxic herbicides such as bromoxynil (which has been 'engineered' into cotton)[9] can have benefits to the environment.

Cotton suffers from enormous weed problems, and is susceptible to all of the broad-leaf herbicides available today. Current weed control in cotton emphasizes massive pre-emergence applications of herbicide, and a mix of hand labor. Field tests conducted by Calgene indicate that by using bromoxynil-tolerant cotton, the yield of cotton can actually be increased while reducing the herbicide usage (and even eliminating pre-emergence herbicides altogether).[9] Of course, since bromoxynil exhibits high toxicity to mammals, it cannot be viewed as the 'environment's friend'. On the other hand, as an alterative to the more long-lasting herbicides currently used with cotton, bromoxynil itself may be an environmental improvement. Thus, the ultimate consequences of bromoxynil-tolerant cotton depend not so much on the recombinant trait itself, but on the agronomic practices associated with usage of this new seed line.

More 'Risky' Applications

In candid conversations, many environmentalists admit that transgenic crops are probably the least worrying of biotechnology's products. One reason is that crops tend to have been debilitated by years of breeding, so much so that outside of cultivated habitats plants such as tomatoes and oilseed rape have little chance of survival. In contrast, much more concern is warranted by transgenic fish, or genetically engineered microbes—organisms that thrive outside the confines of managed habitats. A few illustrative examples will make this clear.

Rapid-growing or 'giant' fish have been engineered by the addition of recombinant genes for growth hormones.[10] While the appeal of these fish for aquaculture is obvious, their escape into natural waters could prove disastrous (especially since existing aquacultural practices with salmon have already had severe impacts on natural fish populations).[11] Even more dubious is the idea of engineering baculoviruses to serve as lethal 'biopesticides' for the control of pest outbreaks. The motivation here is that naturally occurring insect viruses tend to kill too slowly and be too expensive to apply; by constructing a more lethal virus. It might be possible to use viruses to control forest and crop defoliators without introducing any hazardous chemicals to the environment.[12] However, although baculoviruses present no threat to vertebrates, their host range among insects is not well-defined—viruses that currently only infect many insect species incidentally could be turned into lethal agents if they were they engineered to contain strong toxins (since even a few virus particles would be sufficient to kill when toxicity is elevated).

Research and Consensus

Ecologists have begun conducting research that directly addresses the environmental worries discussed above. The most extensive study to date involves a joint industry-government project conducted by a team of researchers headed by Mick Crawley at the Centre for Population Biology at Imperial College in Silwood Park.[13] Crawley and his colleagues grew transgenic oilseed rape in a wide range of habitats. In addition, they manipulated the degree of competition from other plants and levels of attack by fungi, snails, insect herbivores and vertebrate herbivores. Crawley's team found no difference between transgenic rape and unmodified rape.

While this research is noteworthy for its demonstration of an experimental design suited to a controversial question. The British field experiments have done little towards establishing an absence of risk in general. The recombinant trait was herbicide tolerance, yet none of the experimental treatments included herbicide applications—this means there was no opportunity for the modified plant to demonstrate any enhanced vigor. A more pertinent investigation would focus on a transgenic plant in which the modification gave the plant an advantage in a wide variety of circumstances (for example, a plant engineered for insect resistance through the inclusion of a Bt endotoxin gene).

Experiments such as the one pioneered by Crawley's group are now being conducted for a wide variety of crops throughout the world. For example, we are studying the invasiveness of canola whose seed oils have been modified (with potential consequences for germination biology and early seedling growth). Others have conducted field experiments on the invasiveness of hybrids between transgenic canola and weedy mustards,[14] or have sought statistical predictors of weediness in general.[15] One important problem ecologists need to face as they conduct experiments to test the invasiveness of transgenic plants is the issue of statistical power. When asking whether a transgenic plant behaves any differently than parent stock, there is the danger of reporting 'no differences' because the experimental design was ill-conceived or included too few replicates, as opposed to finding 'no differences' as a result of genuine biological similarities.

Risk Analysis, Not Risk-mongering

As ecologists who do research in the areas of environmental toxicology and pest management, we often are bewildered by the debates surrounding biotechnology. On the one hand, Jeremy Rifkin's astounding tales of frightening futures do not mesh with the reality of a tomato modified to ripen better on the vine. On the other hand, industry's tendency to think it fully understands the risks, and smug assurances that genetic engineering is no different to traditional plant breeding do not ring true (especially given our memory of the 'pesticide revolution'). In lieu of ideologically based arguments, ecologists advocate the collection of primary data to evaluate the risks associated with agricultural biotechnology. Moreover, both the British Ecological Society and the Ecological Society of America have developed clear recommendations on how to go about risk assessment regarding genetically engineered organisms.[16,17] Instead of focusing on the technology as the issue, both of these ecological societies favor an emphasis on the product. In other words, the fact that a transgenic tomato proposed for field-testing includes a 'fish gene' is not an issue;[18] the issue is the demographic and ecological performance of the tomato plant that results from the new gene.

Horror tales about genetic engineering emphasize 'monster fantasies', with caricatures like that of the half fish-half tomato. Ecologists realize that if a tomato plant that had fish heads hanging from its branches was produced, it would indeed be quite scary—but the fact that a tomato plant has fish genes incorporated into its genome does not mean the product is anywhere near as frightening. Indeed, fears about 'playing God with Nature' neglect one crucial fact: one third of all existing plant species probably arose from hybridization, that is by moving genes (in fact entire chromosomes) between species. An even more pertinent lesson from evolution is the unity of all life through its DNA. If the human race shares as much as 90% of its DNA with a chimpanzee's DNA, then just what would it mean to move a chimpanzee gene into a human?

In fact, molecular studies have revealed remarkable conservatism of certain genes among species as distantly related as mould and dogs.[19] Obsessive emphasis on between-species movement of genes mistakenly implies some sort of vitalism about genes, as though the fact that a piece of DNA comes from a domestic cat means that any recipient of that DNA obtains some essence of the animal. Ecologists know better. They focus their risk analysis on the product and not the source of the genes. This distinction between the products of biotechnology as opposed to where the genes come from is essential to a constructive debate about biotechnology.

Can Technologies Be Labeled as 'Green' or 'Dirty'?

Field experiments on the invasiveness of transgenic crops are certainly a necessary component of biotechnology development. But we believe such research is inadequate by itself—it is defensive as opposed to forward-looking. One of Rifkin's useful points is that we should think more about the consequences of technologies before we go full speed ahead with their development.[1] He is right, but this vision can be constructive as well as prohibitive. In particular, ecologists could contribute much to biotechnology by designing experiments to identify the types of plant modifications that could lead to more sustainable ecosystems.[20] Thus, instead of simply quantifying how invasive a transgenic cotton might be, we should also be asking what genes might be added to cotton to make cropping systems less environmentally degrading, whether or not the biotechnology revolution repeats the pesticide legacy will have as much to do with the creativity of ecologists as it will with the motives and character of high-technology companies.

References

1 Seabrook. J., *New Yorker*, 19th July 1993
2 Mander, J., *GrantsLists*, 1993, **6,** 1-2
3 Kareiva P., 1993, *Nature*, 1993, **363,** 580-1
4 Pimental, D., et al, *BioScience*, 1989, **39,** 606-14
5 Whitely, H., & Schnepf, H., *Ann. Rev. Microbiol.*, 1986, **40,** 546-9
6 Jepson, P., Croft, B., & Pratt, G., *Molecular Ecology*, 1993, **2,** in press
7 Goldburg, R., Risslen J., Shand, H., & Hassebrook, C., 'Biotechnology's bitter harvest', *Washington, DC: Biotechnology Working Group,* 1990
8 Gianessi, L., 'Biotechnology and US agriculture: the role of herbicide resistant crops', in 'Resources for the future', *Washington: DC,* transcript of seminar, 25th March 1992
9 Calgene, 'Petition for determination of non-regulated status: BXNTM cotton', submitted to the USDA 14th July 1993
10 Fischetti, M., *Science,* 1991, **253,** 512-3
11 Hindar, K., Ryman, N., & Utter, F., *Can. J. Fish Aq. Sci.,* 1991, **48,** 945-57
12 Wood, H., & Granados, R., *Ann. Rev. Microbiol.,* 1991, **45,** 69-87
13 Crawley, M., et al, *Nature,* 1993, **363,** 620-3
14 Linder, R., & Schmitt, J., *Molecular Ecology,* 1994, **3,** 23-30
15 Perrins, J., et al, *Biolog. Conserv.,* 1992, **60,** 47-56
16 Tiedje, J., et al, *Ecology,* 1989, **70,** 298-315
17 Shorrocks, B., & Coates, D., 'The release of genetically-engineered organisms'. *Shrewsbury: British Ecological Society,* 1993
18 USDA APHIS Field Trial Permit, #91-079-01, issued June 1991 (DNA plant technology, thermal hysteresis in tomato)
19 Avers, C., 'Process and pattern in evolution', *Oxford Press,* 1989
20 Rissler, J., & Mellon, M., 'Perils amidst the promise', *Union of concerned scientists,* 1993 ∎

Questions

1. What is the real source of the public's distrust of biotechnology, according to the author?

2. What are the benefits and disadvantages of bromoxynil in cotton growth?

3. What do Bt endotoxins give to plants, and why might the use of endotoxins be detrimental?

Answers begin on page 161.

38 *Biotechnology provides the means by which genetic information (genes) from one organism can be introduced into the genetic material of another organism. Transgenic animals have been genetically modified by the incorporation of genetic information from another species. This gene transfer technology is likely to play an important role in developing animals with enhanced production traits, disease resistance, and the ability to produce valuable drugs. However, there is a good deal of public concern about the health consequences of using biotechnology to create transgenic animals.*

Transgenic Livestock in Agriculture and Medicine

Caird E. Rexroad, Jr.

Although still in its relative infancy, transgenics is already offering a chance to aid organ transplants and produce drugs, as well as enhancing foodstuffs.

The role of livestock in the culture of the post-modern world is being revised. Technology has always contributed to the definition of the interdependent relationship between man and livestock: consider the effects of the wheel, the spinning wheel, the plough, pillows, the milking machine, antiserum production and the freezing of meat. As we begin to understand the genetics underlying the development and function of every cell, and thus develop technologies to manipulate the genetic programme, this relationship will undergo another revolution. Some aspects of the relationship will be extended.

Humans have long used animals as a source of medical compounds, including replacement hormones like insulin. Body parts such as swine skin have also been useful in medicine. In the near future, animals may produce life-saving pharmaceuticals in their milk that would otherwise be impractical to make or isolate. Livestock may be modified so that their organs don't induce immune rejection in humans and can therefore act as life saving transplants. Products from animals, such as milk, meat and wool, may well have their composition or characteristics modified to make them more healthful and useful to man. Man may also be able to contribute to the well being of animals by transferring into them the genetic information for resistance to common diseases. The research that defines these possibilities and enhances the opportunities for livestock husbandry in the future has just begun.

Genetic Manipulation

The basis for modifying the genetic composition is the fact that the genetic code held in the DNA in every cell programmes the events within the animal from conception to death. The genetic programme is carried in two types of DNA: structural genes and regulatory elements. Structural genes code for specific proteins that are the constituents of cells such as enzymes, muscle proteins and hormones. Regulatory elements activate structural genes under the proper conditions and in the appropriate cells of the body. Current recombinant DNA technologies make it possible to build 'designer genes' by making new combinations of regulatory elements and structural genes.

Another important basis for modifying the genetic programme of livestock is the recognition that parts of a new genetic programme can be inserted into animals in order to reprogram them. When DNA containing genetic information is injected into fertilized eggs the result has been a minor modification of the genetic programme in all mammals studied so far, including mice, rats, rabbits and livestock species.[1,2] These 'transgenic' animals carry new genetic information and can pass it on to their offspring. The group of scientists in the Institute of Animal Physiology and Genetics Research at Roslin led by A. J. Clark mixed gene insertion with recombinant DNA technologies to produce pharmaceuticals in milk.[3] They combined a regulatory ele-

Reprinted with permission from *Chemistry & Industry*, May 15, 1995.

ment, a promoter, from the sheep beta lactoglobin gene with the structural gene for human α_1-antitrypsin. This new gene resulted in human α_1-antitrypsin being made in the sheep's milk.

Since the first genetically engineered livestock were reported in 1985,[2] over 40 studies have been conducted to insert new genes into livestock species.[4] The introduction of new genes into livestock is done by microinjecting DNA into the nuclei of newly fertilized eggs recovered surgically from the oviducts of sheep, cows and pigs, or by microinjecting cow embryos produced completely in the laboratory starting with slaughterhouse ovaries and commercial semen. The usual procedure includes hormonal induction of superovulation of donor females, embryo recovery, embryo microinjection, embryo culture and embryo transfer. Offspring resulting from these microinjected eggs only rarely incorporate the new genetic information: the rates based on microinjected eggs transferred to surrogate mothers are about 0.9% for livestock species.[4] Such low rates are problematic for performing a large number of studies in livestock, mainly because of the cost incurred in keeping surrogate mothers to carry the microinjected eggs to term. The costs are most easily justified for the production of valuable pharmaceuticals.

A Growth Area
The first transgenic livestock were reported by a collaboration that included Ralph Brinster, Robert Hammer and Karl Ebert at the University of Pennsylvania, myself, Vernon Pursel, Robert Wall and Douglas Bolt at the Agricultural Research Service, Beltsville, and Richard Palmiter at the University of Washington, Seattle.[2] Soon after a group led by Gottfried Brem at Munich reported similar transgenic studies.[13] These studies were aimed at improving the growth characteristics of pigs and sheep. The first studies tried to enhance animal growth by inserting either growth hormone genes or genes to release growth hormone from the animals' own pituitary. This work depended on recombinant genes in which the growth hormone gene or growth hormone releasing factor structural gene were regulated by 'hot' promoter which activates transgenes in large organs, such as the liver, thus elevating growth hormone levels in the blood stream. The studies were both highly successful and sobering.

The transgenes, frequently made of mouse regulatory elements and human, pig or cow structural genes, functioned in the transgenic pigs or sheep much as expected, elevating blood growth hormone levels. In addition, some of the expected physiological reactions were observed with transgenic pigs having more rapid growth and reduced food consumption. Pigs with elevated growth hormone also had reduced fat in their carcass.[4] The sobering aspect was that these first transgenes were not well controlled, they only worked by making excessive amounts of product. The result was that the transgenic pigs had joint problems, ulcers and were infertile. The transgenic sheep developed a condition similar to diabetes.

More recent approaches, while still not fully successful, have used regulatory elements that rely on the more subtle regulation of transgenes. For example, by having the transgene activated in response to feeding in pigs,[5] or restricting the response to a particular tissue, such as muscle, that is then stimulated.[6] Additional studies are looking at stimulating not growth hormone but specific genes that affect the cascade of events established by a rise in growth hormone levels. The successful use of growth hormone to increase productivity in pigs by serial injection of the recombinant protein suggests that the transgenic approach still has potential, but it relies on the development of genes that can be finely regulated.

Transgenics and Milk
The single most studied transgene system in livestock species seems to be one to modify the peptides produced in the mammary gland. In the US, dairy cows make over 3bn kg of protein annually.[7] This rich protein source could provide a wealth of proteinaceous pharmaceuticals ('gene pharming') or milk with enhanced health value ('nutraceuticals') or simply milk that produces better foods (such as high casein for cheese production).[8]

Many human diseases are associated with a mutation in a gene that codes for an important protein such as a blood clotting factor. The usual treatment is to replace these proteins with cadaver or blood-derived proteins. Transgenic animals could be a rich source of these proteins. The group at Roslin has produced sheep carrying transgenes that make the human protein, blood clotting factor IX, in milk.[9] This protein might be useful for treating some forms of hemophilia.

Since the cells that make the proteins are mammalian, the proteins that are produced are in many ways processed as they would be in humans. Their presence in milk makes them readily obtainable and perhaps relatively easy to prepare. These transgenic 'factories' are really reproduced and studies have suggested that the transgenes are, for the most part, stable from generation to generation. Production in livestock avoids problems associated with contamination of human diseases such as the HIV retrovirus, which in the past has contaminated human blood sources. On the other hand, producing medicines in livestock is not completely without concern and any such work would have to ensure that the animals involved were disease free.

A consortium of researchers from the University of Leiden and the company Gene Pharming Europe BV has investigated the possibility of improving milk quality for consumption by human babies.[10]

The researchers produced the bull 'Hermann', which transmits a gene to his daughters to make them produce human lactoferrin in their milk.[10] Lactoferrin, which is relatively low in cows' milk and high in human breast milk, may improve the transfer of iron to babies and reduce the possibility of bacterial infection. These studies have raised concern about the ethics of using livestock in this way and may not be followed to completion.

Combatting Disease
Many disease organisms exploit single 'weak links' when they attack their prospective host. The Visna virus, a retrovirus similar to the human immunodeficiency virus, attacks the immune system by attaching to specific sites (receptors) on the surface of cells. A possibility being investigated for confounding this retrovirus is to fill those receptor sites with part of the virus' coat protein so that the whole infectious virus cannot then bind to the cell. This approach has been used in chickens by infecting them with a defective retrovirus and has been successful in lowering infection levels after retroviral exposure.[11]

We have produced transgenic sheep using a transgene made up of the regulatory region of the retrovirus and the coding region for the coat (envelope) protein.[12] Three ewes, microinjected as embryos, produced the coat protein in the cells (macrophages) of their immune system—these cells are normally the targets of retroviral infection—and not in other cells of the lineage that produce blood white cells. Satisfactory testing of these animals for disease resistance is waiting for a sufficient number of transgenic offspring to be produced. One interesting finding, however, is that the sheep produced antibodies to the coat protein. This may have been caused because the coat proteins are made in cells that are part of the body's response to infection, so the proteins are recognized by the body as being foreign. Whatever the reason, antibodies produced in response to the coat protein would suggest that genetic immunization via the transgene approach is a possibility.

The most exquisite genetic engineering experiments are those in each individual genes have either been replaced or 'knocked out'. These experiments in mice have relied on the availability of embryonic stem cells, cells which have the potential to contribute to every cell in the body. Typically, these cells are grown in culture where the genes are inserted by one of several techniques. After gene insertion, the cells that have been appropriately manipulated (perhaps as few as one in 50,000) are then selectively cultured. These manipulated cells are then transferred into early embryos where they may become part of the fetus and hopefully part of the germ line so that the genetic information can be passed on to their offspring. Some of these stages have been realized in livestock. Embryonic stem cells have been reported by a number and at least one group has shown that offspring can be produced from the cultured cells.

The Future
Societal concerns and scientific problems are currently barriers to the introduction of genes into livestock. The scientific problems relate to lack of knowledge in three areas: the identity of the specific gene(s) that produce desirable traits, gene regulation, and the mechanisms of DNA insertion into the genome. Gene mapping research currently underway in livestock species may identify beneficial variations of genes that already exist. However, not all genes that are to be used would necessarily pre-exist in Nature. Gene regulation is the subject of many fundamental studies and the results have already begun to enhance work on transgenic livestock. Minimal research is being conducted on the incorporation of genes into the genome and more needs to be accomplished.

Societal concerns about the safety of new technology that the general public poorly understands may limit research: much of the world population perceives that an ample food supply is already available. This short-sighted attitude does not address the changes that will occur in world populations and climate conditions in the next 50 years. Research in animal agriculture will be underfunded unless this attitude changes. A different kind of societal problem is the acceptance of the use of transgenic animals. This subject is one of continued debate. The research and industrial community needs to present the objective findings of science so that the debate has more than emotional substance. The use of livestock as organ donors for human needs may help redefine the general view of the use of transgenic animals.

References
1. Gordon, J.W., & Ruddle, F.H., *Science*, 1981, **214**, 1244-6
2. Hammer, R.E., *et al*, *Nature*, 1985, **315**, 680-3
3. Wright, G., *et al*, *Bio/Technology*, 1991, **9**, 830-4
4. Pursel, V.G., & Rexroad Jr., C.E., *J. Anim. Sci.*, 1993, **71(3)**, 10-19
5. Wieghart, M., *et al*, *J. Reprod. Fert. Suppl.* 1990, **41**, 89-96
6. Hill, K.G., *et al*, *Theriogenology*, 1992, **37**, 222
7. USDA, 'Agriculture Fact Book', 1994
8. Wilmut, I., *et al*, *J. Reprod. Fert.*, 1990, Suppl. 41, 135-46
9. Clark. A.J., *et al*, *Bio/Technology*, 1990
10. Krimpenfort, P., *et al*, *Bio/Technology*, 1991, **9**, 844-7
11. Crittenden, L.B., & Salter, D.W., *J. Reprod. Fert.*, 1990, Suppl. 41, 163-71
12. Clements, J.E., *et al*, *Virology*, 1994, **200**, 370-80

13. Brem, G., *et al, Zuchthygiene,* 1985, **20,** 251-2
14. Pursel, V.G., *et al, Vet. Immunol. Immunopathol.,* 1987, **17,** 303-12
15. Vize, P.D., *et al, J. Cell. Sci.,* 1988, **90,** 295-300
16. Brem, G., *et al, Occasional Publ. Brit. Soc. An. Prod.,* 1988, **12,** 15
17. Ebert, K.M., *et al, Mol. Endocrinol.,* 1988, **2,** 277-83
18. Pursel, V.G., *et al, Science,* 1989, **244,** 1281-8
19. Roushlau, K., *et al, J. Reprod. Fert. (Suppl.),* 1989, **38,** 153-60
20. Rexroad, Jr, C.E., *et al, Molec. Reprod. Dev.,* 1989, **1,** 164-9
21. Murray, J.D., *et al, Reprod. Fertil. Dev.,* 1989, **1,** 147-55
22. Polge, E.J.C., *et al,* in: 'Biotechnology of growth regulation', (Eds R.B. Heap, C.G. Prosser & G.E. Lamming), 1989, 189-99
23. Ebert, K.M., *et al, Anim. Biotech.,* 1990, **1,** 145-59
24. Weighart, M., *et al, J. Reprod. Fert. (Suppl.),* 1990, **41,** 89-96
25. Massey, J.M., *ibid,* 1991, **42,** 199-208
26. Rexroad Jr, C.E., *et al, J. Anim. Sci.,* 1991, **69,** 2995-3004
27. Pursel, V.G., *et al, Thieriogenology,* 1992, **37,** 278
28. Lo, D., *et al, Eur. J. Immunol.,* 1991, **21,** 1001-6
29. Weidle, U.H., *et al, Gene,* 1991, **98,** 185-91
30. Brem, G., *Mol. Reprod. Dev.,* 1993, **36,** 242-4
31. Simons, J.P., *et al, Bio/Technology,* 1988, **6,** 179-83
32. Clark, A.J., *et al, ibid,* 1989, **7,** 487-92
33. Ebert, K.M., *et al, ibid,* 1991, **9,** 835-8
34. Wall, R.J., *et al, Proc. Natl. Acad. Sci. USA,* 1991, **88,** 1696-1700
35. Velander, W.H., *et al, ibid,* 1992, **89,** 12003
36. Ward, K.A., & Nancarrow, C.D., *Experientia,* 1991, **47,** 913-22
37. Rogers, G.E., *Trends Biotech.,* 1990, **8,** 6-11
38. Bawden, C.S., *et al, Transgenic Research,* 1995, **4,** 87-104
39. Bullock, D.W., *et al,* in 'Proceedings of Beltsville Symposium XX', 1995, 8 ■

Questions

1. What are the combinations that comprise "designer genes"?

2. What factor makes studying transgenic animals so expensive?

3. What problem is avoided if animals make transgenes?

Answers begin on page 161.

39 *Parents have long preached the virtues of spinach, carrots, and other garden produce. Now there's an explosion of compelling studies associating diets rich in fruits and vegetables with a lower cancer risk. Researchers are discovering a number of natural substances called* functional components *in fruits and vegetables that appear to hinder tumor growth. The identification of these active substances and the discovery of how they interfere with carcinogenesis are an extremely active area of research, with new reports coming out almost daily.*

Cancer-fighting Foods: Green Revolution

Kristine Napier

No one wants to be one of the 1.2 million Americans diagnosed with cancer each year. In an effort to avoid this all too common fate, people may fill up on fiber, obsess about antioxidants, or shun red meat and fat. In recent years, however, scientists have realized that these dietary elements are only the tip of the iceberg when it comes to reducing cancer risk. A previously hidden world of natural chemicals in edible plants is unfolding, and the more researchers learn, the more certain they are that mom was right; we should eat our vegetables, and lots of them.

"There's an explosion of compelling and consistent data associating diets rich in fruits and vegetables with a lower cancer risk," said epidemiologist Tim Byers, who studies the relationship between diet and chronic disease at the Centers for Disease Control and Prevention in Atlanta. One analysis of data from 23 epidemiologic studies found that a diet rich in vegetables and grains slashed colon cancer risk by 40%. Another study found that women who ate few vegetables had an incidence of breast cancer that was about 25% higher than those who consumed more produce. All in all, at least 200 epidemiologic studies from around the world have found a link between a plant-rich diet and a lower risk for many types of tumors.

Findings such as these have inspired laboratory scientists to try and analyze just what it is about fruits and vegetables that might fend off cancer. "There's more to food than vitamins, minerals, fiber, calories, and protein," said cancer epidemiologist John D. Potter of Seattle's Fred Hutchinson Cancer Research Center. "We're discovering a plethora of bioactive substances in plant foods." Called *functional components,* these include a large class of naturally occurring compounds known as phytochemicals. Meanwhile, "many traditional nutrients, including folic acid and selenium, have functions that are becoming clearer—including an ability to fight cancer," said Dr. Potter.

This powerful epidemiological evidence is being bolstered by newer laboratory studies showing how functional components interfere with carcinogenesis. "These compounds seem to interact with every step in the cancer process, mostly slowing, stopping, or reversing them," Dr. Potter said. Most functional components appear to boost the production or activity of enzymes that act as:
• blocking agents, detoxifying carcinogens or keeping them from reaching or penetrating cells, or
• suppressing agents, restraining malignant changes in cells that have been exposed to carcinogens.

Hotter Than the Internet?
Anyone who hasn't yet heard about functional foods soon will—the term is well on its way to becoming the latest nutrition buzzword. Before settling on this appellation, researchers tossed around at least 20 different names, including "designer foods," "nutriceuticals," "pharmafoods," and "chemo-preventers." But the functional foods label won out in 1994 when it was endorsed by the food and nutrition board of the Institute of

---

**Reprinted from the April 1995 supplement issue of the *Harvard Health Letter*,
© 1995, President and Fellows of Harvard College.**

Medicine. It simply means "foods with ingredients thought to prevent disease."

Although the conventional wisdom is that new trends take hold first on the coasts, in this case a Midwestern institution appears to be out in front. The University of Illinois has the nation's first (and only) full-scale scientific program devoted to the study of phytochemicals and other functional components. The Functional Foods for Health Program (FFH) involves 63 faculty members from more than 20 disciplines and represents both the Chicago and Urbana-Champaign campuses of the university.

"The program combines expertise from agriculture and medicine to study how naturally occurring components in foods may protect people from disease," said FFH director Clare Hasler. In related work, the department of medicinal chemistry and pharmacognosy on the Chicago campus maintains the world's largest database on the chemical constituents and pharmacology of plant extracts.

Phytochemicals
When life began, plants were *anaerobic*—they lived in a world devoid of oxygen. As they evolved and began turning carbon dioxide into oxygen, however, they gradually polluted their own environment. In order to survive, plants were forced to develop defenses against unstable forms of oxygen, explained researcher David Heber, head of the clinical nutrition research unit at the University of California-Los Angeles.

Phytochemicals, many of which are brightly colored and help give plants their vivid hues, are key parts of this antioxidant defense system. In addition to resisting oxidation, these substances guard against an array of adversities including viral attack, harsh weather, and the insults of handling. Eons later, it appears that humans can now benefit from eating plants that contain these disease-fighting substances. Unlike other minor constituents in food, however, phytochemicals have no calories and no known nutritional value. In other words, they are not necessary for normal physiologic function.

There are literally hundreds of phytochemicals, only a sprinkling of which have been studied. They can be categorized in several ways: by chemical name, by primary food source, and by anti-cancer action. Many foods contain numerous phytochemicals, each acting via one or several mechanisms. And because new data are being published almost daily, constant updating is needed to keep any list of phytochemicals current. In this section, they are organized by chemical name.

As exciting as scientists find this area of inquiry, they don't pretend to have all the answers yet. "While there's no doubt that diets rich in fruits and vegetables are cancer-protective, much of what we know about individual phytochemicals is still speculative," cautioned nutrition researcher Phyllis Bowen, co-director of the University of Illinois' functional foods program.

Flavonoids are an array of chemicals widely found in fruits, vegetables, and wine. They may reduce cancer risk by acting as antioxidants: blocking the access of carcinogens to cells, suppressing malignant changes in cells, or a combination of these. "Flavonoids may also interfere with the binding of hormones to cells and thus may inhibit cancer development," said nutrition researcher Diane F. Birt of the University of Nebraska Medical Center's Eppley Institute.

Indoles and **isothiocyanates** (also called mustard oils) are largely responsible for putting broccoli on the cancer prevention map. Both account for some of the "bite" in the taste of cruciferous vegetables, and both are breakdown products of complex plant compounds called glucosinolates. They are formed when these compounds are altered by processing, cooking, or chewing. Scientists believe that indoles and isothiocyanates act mainly by blocking cancer-causing substances before they reach their cellular targets. Isothiocyanates may also suppress tumor growth.

Isoflavones are prominent in soy beans and everything that's made from them. Some scientists believe that differences in soy consumption explain why the incidence of breast cancer in Asian women is 5-8 times lower than in American women, as well as why prostate cancer is lower in Asian men. (See "Diet and the Prostate," *Harvard Health Letter,* July 1994.) Isoflavones can act as antioxidants, carcinogen blockers, or tumor suppressors. Plants contain many forms, including genistein, biochanin A, and daidzein.

Other cruciferous chemicals that are thought to have anticancer properties include dithiolthiones, chlorophyllin (chlorophyll combined with sodium and copper), and organonitriles. In addition to these functional components, cruciferous vegetables are also rich in fiber and in vitamin C and selenium. "No doubt these substances work in some complex synergy to fight cancer," said Matthew A. Wallig, an investigator in functional-foods research at the University of Illinois.

Lignans occur in many foods, but are especially concentrated in linseed (which are seeds from flax, the same plant that is woven into linen cloth). Lignans may have an antioxidant effect and may block or suppress cancerous changes. "Although flax hasn't been used much in this country, an increasing number of health food stores and bakeries are adding it to bread products," said Dr. Bowen of the University of Illinois. This practice started because flax is also high in omega-3 fatty acids, which are thought to protect against colon cancer and heart disease.

Organosulfur compounds are found in plants from the genus *Allium* which includes garlic, onions, leeks, and shallots. Diallyl disulfide is the most potent of these chemicals, which may act as blocking or suppressing agents.

Monoterpenes occur naturally in citrus fruits (one variety is D-limonene) and in caraway seeds (in the form of D-carvone). Scientists think they act by interfering with the action of carcinogens.

Saponins are a large family of modified carbohydrates found in many vegetables and herbs. So far, researchers have identified 11 different saponins in soybeans alone. In addition to having anti-cancer activity, there is evidence that some of these substances break down red blood cells, deactivate sperm, or lower circulating levels of certain lipids.

Carotenoids
Although phytochemicals have been hogging the spotlight of late, the red and yellow plant pigments known collectively as *carotenoids* are also thought to be potent cancer fighters. This is still true, even though beta carotene supplements lost some of their luster after several large studies failed to demonstrate the kind of anticancer activity that many people had hoped for. (See "Second Thoughts About Antioxidants," *Harvard Health Letter*, February 1995.)

Researchers may have been too quick to assume that beta carotene by itself deserved credit for lower cancer rates, according to nutritional epidemiologist Regina G. Ziegler of the National Cancer Institute. People who ate diets rich in fruits and vegetables or who had high circulating levels of beta carotene showed a reduced risk of cancer. But, she pointed out, "blood levels of beta carotene may simply be a good marker for fruit and vegetable intake."

Beta carotene may be beneficial in its natural form, bound up with other constituents of food, but not when it is isolated as a supplement. It is also possible that other carotenoids may be the real cancer inhibitors, and that they may be more efficacious against some types of carcinogens and tumors than others.

In addition to having antioxidant properties, carotenoids may work in several other ways, Dr. Bowen said. They may enhance normal communication among healthy cells, a buzz of biochemical conversation that scientists think helps keep cancer cells from running amok. It's also possible that beta carotene is transformed into retinoic acid, a substance that some researchers say can turn on and off genes that may play a role in cancer development.

New Roles for Old Nutrients
The phytochemicals and carotenoids are just two dietary defenses in the war against cancer. As scientists learn more about how cancer progresses, they are finding that some traditional vitamins and minerals also show protective promise.

Folate (also known as folic acid) is best known for its role in the formation of healthy red blood cells. Now there's compelling epidemiologic evidence that people with higher folic acid levels are less likely than others to develop colon cancer and precancerous colon polyps, according to researcher Joel B. Mason, an assistant professor of medicine and nutrition at Tufts University. "This relationship has been uncovered just in the last few years," said Dr. Mason, who is coordinating a multicenter trial probing folic acid's ability to modify colon cancer risk.

Other researchers have shown that folate contributes to normal tissue formation by guarding the integrity of the genetic messages encoded in DNA, and Dr. Mason speculates that this protective effect may thwart carcinogens that would ordinarily cause colon cancer. But he isn't ready to recommend that people who are worried about colon cancer take large doses of folic acid. "It's definitely a good idea, though, to get the RDA of 200 micrograms (mcg), or perhaps up to 400 mcg, the level to which some experts recommend raising it." In his experiments, Dr. Mason uses doses 20 to 40 times greater than the RDA, which would not be safe for everyone. High folate intake can increase seizure risk for people with epilepsy, for example, or may mask B^{12} deficiency, which can lead to serious neurological troubles, especially in older people.

Calcium appears to have some preventive value where colon cancer is concerned. Researchers propose several mechanisms to explain how calcium acts as an anti-cancer warrior in the colon; these include inhibiting cell growth and/or disarming potential toxins by binding them to fatty acids.

Selenium is being actively studied by epidemiologists and basic scientists with mixed results. Interest was sparked by epidemiologic evidence that population groups with higher selenium intakes have less cancer than those who consume little of this trace mineral.

Although these findings were supported by results from animal experiments, further epidemiologic investigations have found little or no protective effect in humans. Some researchers believe that selenium may work best as an anti-cancer agent in concert with phytochemicals or antioxidants such as beta carotene and vitamin C. A large scale clinical trial now underway is expected eventually to shed more light on this mineral's possible benefits.

Conjugated linoleic acid's beneficial effects suggest that people cannot live by vegetables alone, and that a more varied diet may be best. "Substances that fight cancer may not always be in plants," said food and nutrition researcher Michael W. Pariza of the Food Research Institute at the University of Wisconsin-Madison.

In the 1970s, Dr. Pariza and his co-workers were investigating possible carcinogens formed by grilling meat when they stumbled across a substance that appeared to inhibit cancer instead of contributing to it. This was conjugated linoleic acid, a form of an essential fatty acid which is found in beef and in fat-containing dairy products.

When Dr. Pariza tested the substance in animals that spontaneously develop breast cancer, he found that the conjugated acid slows the growth of cells that give rise to cancer. There is some laboratory evidence that it does this by jump-starting the immune system, which repels cancerous changes. This does not mean that people should "chow down on dairy fat," he emphasized. But it does suggest that moderate consumption may be better than none at all.

Vitamin A is what the body produces when it metabolizes carotenoids; it's also found in dairy products and animal fat. Some studies indicate that vitamin A itself, either from food or supplements, may also offer some cancer protection.

Vitamin D's role is unsettled right now; early studies indicated that it might provide some protection against colon cancer, but subsequent ones weren't as promising.

But What to Eat?

Will Americans soon be slurping down an elixir of lignans, flavonoids, saponins, and folic acid? Most likely not, agree the experts consulted for this article. "I don't condone emphasizing one or even several functional components," said Dr. Hasler from the University of Illinois. "Phytochemicals and other dietary substances no doubt work in concert to fight cancer and other diseases. In addition, isolated phytochemicals may actually be harmful at high doses."

"Simply put," said nutrition and cancer expert Cheryl Rock, an assistant professor at the University of Michigan, "the best advice is to eat real food instead of relying on supplements. If you just take supplements, you simply don't get all of the compounds in foods we're still learning about. We don't know yet if we should combine an indole with an isoflavone, or folic acid with selenium. Right now, only nature knows best."

Recommendations from the National Cancer Institute (NCI) and others emphasize the importance of eating at least five to nine servings of fruits and vegetables each day, aiming for a wide variety. This isn't as hard as it seems: try for one or two fruits at breakfast, one fruit and two vegetables at lunch and dinner, and a snack to make a total of nine.

Although the NCI doesn't fine-tune its advice about exactly what to eat, many experts believe that people can best attain a balance of beneficial substances by making sure that their diet includes foods from each of the following categories:
- cruciferous vegetables
- citrus fruits
- dark green leafy vegetables
- dark yellow/orange/red vegetables

A good rule of thumb is to eat at least three different colors of fruits and vegetables every day. "We know, for example, that the red pigment in tomatoes has completely different bioactive ingredients than the orange pigment in carrots; the same is true for the bioactive ingredients in citrus fruits versus those in the cruciferous vegetables," said Dr. Hasler.

And be sure to eat other plant foods as well, said Dr. Potter. "Grains, nuts, seeds, and legumes also contain a wide variety of bioactive compounds."

Humans got their start as gatherers, probably eating little bits of many different fruits and vegetables every day, and we should strive for such variety and quantity again, said Dr. Potter. "Vegetables and fruits contain the anticarcinogenic cocktail to which we are adapted. We abandon it at our peril," he said. ■

Questions

1. What are functional foods?

2. How do phytochemicals act as a defense in plants?

3. In what food source are isoflavones found, and how do they help fight cancer?

Answers begin on page 161.

40 *The beneficial health effects of the B vitamins have been documented for some time. For example, a lack of vitamin B6 can result in skin lesions and in disturbances in the central nervous system. A vitamin B_{12} deficiency is characterized by large immature red blood cells and a progressive paralysis of the nerves. Folic acid prevents a number of specific anemias, such as those associated with pregnancy and malnutrition. Growing evidence now suggests that B vitamins may be far more important than was previously thought.*

B Makes the Grade

A flurry of recent studies is propelling the once-lowly B vitamins toward the head of the class.

In 1989, researchers thought so little of folic acid that the National Academy of Sciences cut in half its recommended daily intake of that B vitamin. Now the U.S. Food and Drug Administration is seriously thinking about requiring food manufacturers to fortify refined-grain products with folic acid.

That striking reversal in folic acid's fortunes stems from studies showing that the vitamin can sharply reduce a woman's risk of giving birth to a child with either spina bifida or anencephaly, two devastating neurologic defects. But folic acid—and two other long-neglected B vitamins—may have health benefits that go far beyond the prevention of birth defects. Growing evidence now suggests that B vitamins may help keep the mind and the immune system strong, and they may reduce the risk of deadly disease.

Cardiovascular Disease
Researchers have long known about a rare genetic disease that causes extremely high blood levels of the amino acid homocysteine. Most victims die in childhood of heart attacks, because homocysteine can damage the arteries and may thicken the blood.

Starting in 1990, a stream of studies has transformed high homocysteine levels from a rare genetic threat to a major public-health concern. The first of those reports, including a large study from Harvard University, showed that even a moderately elevated level of homocysteine in adults increases the risk of heart attack. Scientists already knew that homocysteine levels are controlled by several ordinary B vitamins. So in theory, failure to get enough of the vitamins could push those levels to dangerous heights.

The following year, researchers from Tufts University, using data from the ongoing Framingham Heart Study, confirmed that people who consume too little folic acid do indeed have high homocysteine levels. On average, only those who consumed at least 400 micrograms per day (ironically, the amount formerly recommended by the National Academy of Sciences) had homocysteine levels low enough to pose no apparent threat to the heart. There was a weaker but still significant connection between a shortage of either vitamin B6 or B12 and excessive homocysteine.

Meanwhile, South African researchers showed that giving volunteers a combination of all three vitamins can reduce homocysteine levels by 60 percent.

In 1994, a study from the University of Alabama finally linked one of the B vitamins directly with coronary risk. The researchers evaluated about 100 men with coronary heart disease and another 100 men without coronary disease. Those in the top quarter for blood levels of folic acid were nearly 40 percent less likely to have the disease than those in the bottom quarter. Then a larger study from Tufts extended that finding to a major risk factor for stroke. Individuals who consumed only moderate amounts of folic acid were about 50 percent more likely to have dangerous clogging of the carotid artery—the main vessel feeding the brain—than those who consumed large amounts of the vitamin.

The final, most crucial step—clinical trials that would prove or disprove the link between B vitamins and reduced coronary risk—has not yet been done.

Cancer
Folic acid and vitamin B12 help the body convert homocysteine to methionine; that amino acid helps

"B Makes the Grade." Copyright 1995 by Consumers Union of U.S., Inc., Yonkers, NY 10703-1057. Reprinted by permission from *Consumer Reports on Health*, June 1995.

maintain the genes that prevent cells from turning cancerous. In theory, an inadequate supply of those vitamins will reduce methionine levels and, in turn, increase the risk of cancer.

The few studies done so far tend to support that theory, at least for folic acid. In the largest study, Harvard researchers evaluated the folic-acid intake of some 25,000 people. Those who consumed the most folic acid were about one-third less likely to have precancerous colon polyps than those who consumed the least. Two smaller studies have linked high consumption of folic acid from food with a 40 to 70 percent reduction in the risk of colorectal cancer itself. And preliminary reports suggest that even a marginally low intake of the vitamin increases the chance of precancerous changes in at least two other sites: the cervix and the lungs. Again, no clinical trials have tested the link between folic acid and cancer.

Mind and Mood
The body needs folic acid, B6, and B12 to manufacture neurotransmitters, chemicals that control alertness and mood by speeding nerve signals through the brain. Studies have found that some people who are seriously depressed or show signs of dementia have deficiencies of folic acid or B12. Correcting those deficiencies can ease the depression or improve concentration and memory.

More important for the average person, even a mild lack of the B vitamins may cloud the mind. In one study of 260 older people who showed no signs of illness or vitamin deficiency, those with the lowest blood levels of B12 and folic acid scored significantly worse on tests of mental acuity than the rest of the group did. In a second study, Tufts researchers artificially created a deficiency of vitamin B6 by feeding older volunteers a specially restricted diet; modest supplements then corrected the deficiency. Memory deteriorated steadily as B6 levels fell, and returned to normal when adequate levels were restored.

Immunity
A sufficient supply of vitamin B6 and, to a lesser extent, of folic acid is essential for keeping the immune system strong. Even people with only marginally low levels of those vitamins show signs of weakened immune function, mainly reductions in the number and activity of certain disease-fighting white blood cells. In one clinical trial involving older people who had several modest vitamin deficiencies, low B6 levels at the start of the study were strongly correlated with reduced immune response. One year later, the volunteers who received a multivitamin/multimineral supplement which corrected the deficiencies, had significantly stronger immune-cell activity than those who received a placebo.

Some evidence suggests that just getting enough B6 to prevent outright deficiency may not be enough to maintain optimal immune strength. In the Tufts study of B6 and memory described above, the researchers also evaluated immune function. Like memory, immune response deteriorated as B6 levels declined, and it returned to normal when the deficiency was eliminated. Toward the end of the study, however, the volunteers received a large dose of B6, about 25 times the recommended intake, just to be sure the deficiency had been corrected. That precaution unexpectedly boosted immune response to a point well above the original B6 levels, closer to the levels seen in younger people.

Whether the apparent link between B vitamins and immune-cell activity actually affects the risk of disease is not yet known.

How to Get What You Need
Overall, the research indicates that B vitamins may be far more important than was previously thought. Unfortunately, many people consume an inadequate amount of two of those vitamins. The average daily intake of folic acid by adults is about 285 micrograms, substantially less than the minimum of 400 micrograms formerly recommended by the National Academy of Sciences and currently listed by the U.S. Government as the Recommended Daily Value. And more than half of all American adults get less than the recommended amount of vitamin B6, which is 2 milligrams.

Fortunately, the diet widely recommended by public health authorities—which includes at least five servings per day of fruits or vegetables and at least six servings of whole grains or beans—will typically supply enough folic acid and more than enough B6. Good sources of folic acid include avocados, green leafy vegetables, citrus fruits, beans, sunflower seeds, and wheat germ. Foods rich in B6 include chicken, fish, avocados, potatoes, and watermelon.

There's not much evidence that consuming higher doses of the B vitamins than you'd get from food can significantly increase their benefits. Further, relying on supplements rather than making the effort to eat a nutrient-rich diet deprives you of the many health benefits of such a diet. However, some older people can't eat an adequate diet, much less an optimal one, typically because of dental problems or restricted mobility. They should consider taking a low-dose multivitamin/multimineral supplement.

People who eat meat, fish, or dairy products generally have little trouble getting enough vitamin B12. Even strict vegetarians, who eat no meat or dairy, may get enough B12 if they consume a lot of fortified cereals, soy beverages, or certain types of nutritional yeast—check the label. (Note that fermented foods such as hot-pepper sauce, soy sauce, and tempeh, a soybean product, also contain B12, but most of it is biologically inactive.) Strict vegetarians who don't consume good sources of B12 should consider eating a cereal fortified with 100 percent of the Daily Value for the vitamin or taking a low-dose supplement.

The B Stings

Although recent evidence points to a heightened role for B vitamins in protecting against certain diseases, they are hardly the cure-all implied by the promotion of B-vitamin supplements for stress, fatigue, premenstrual syndrome, carpal tunnel syndrome, and even hangovers.

Stress. Supplements such as *Schiff's High Potency Mega-Stress* and Lederle Laboratories' *Stresstabs* contain hefty doses of vitamins B1, B2, B6, and B12. But emotional stress does not heighten the body's need for B vitamins—or for any other vitamin for that matter. While extreme physical stress, such as running a marathon, may slightly increase the need for B1, B2, and B6, even marathoners can almost always get all the nutrients they need simply by eating a nutritious diet.

Fatigue. Since at least the 1950s, some doctors have injected tired patients with vitamin B12 to boost their energy levels, though there's never been any evidence to support the efficacy of such B12 injections. But lack of evidence hasn't stopped Nature's Bounty, one of the country's largest dietary-supplement manufacturers, from capitalizing on the image of vitamin B12 as an energy booster. It sells a nasally administered vitamin supplement called *Ener-B*, which promises that squirting the gel up your nose, at $1 a snort, will supply a "burst of B12" that "you will feel good about".

The package insert claims that many people lack a protein in the stomach, called intrinsic factor, that the body needs to absorb B12. In reality, lack of intrinsic factor is uncommon. Those who do lack the protein have a serious problem that requires monthly injections of B12, not self-administered snorts of the vitamin.

Carpal-tunnel and premenstrual syndromes. Certain self-help books and even some doctors recommend doses of B6 several hundred times the recommended intake to treat carpal-tunnel syndrome—painful inflammation of the wrist caused by repetitive tasks such as typing—and premenstrual syndrome. But there's little evidence that the vitamin helps either problem. And taking such huge doses of B6 for prolonged periods can damage the nerves, triggering the same sort of tingling and numbness caused by carpal-tunnel syndrome itself.

Hangovers. Some supplements, such as *Source Natural's Hangover Formula*, include megadoses of several B vitamins as a supposed cure for morning-after misery. But popping B vitamins won't make bingers feel any better—it will only turn their urine bright yellow. ■

Questions

1. What does folic acid prevent?

2. What risks are associated with high homocysteine levels, and how do B vitamins affect this amino acid?

3. Why can the intake of folic acid, B_6, and B_{12} affect alertness and mood?

Answers begin on page 161.

41 *Glucose is by far the most plentiful monosaccharide found in blood, and the term* blood sugar *refers to glucose. In adults, the normal blood-sugar level measured after a fast of 8–12 hours is in the range of 70–110 mg/100 mL. The blood-sugar level reaches a maximum of approximately 140–160 mg/100 mL about 1 hour after a carbohydrate-containing meal. It returns to normal after 2.0–2.5 hours. If blood sugar is below the normal level, a condition called* hypoglycemia *exists.*

Hypoglycemia: Fact and Fiction about Blood Sugar

Jamie Spencer

For some people, the classic continental breakfast of a sweet roll, orange juice, and coffee is a recipe for distress. Within a few hours after eating, they may feel light-headed, shaky, and unable to concentrate. Are such people suffering from *hypoglycemia,* a fancy word that simply means "low blood sugar"? Probably not, say researchers who have studied this controversial disorder over the past two decades. Far fewer people have the condition than they imagine.

In the 1970s and early 1980s, hypoglycemia was a fashionable diagnosis for patients with vague symptoms ranging from fatigue to palpitations. Some people theorized that low blood sugar accounted for such disparate social ills as juvenile delinquency and alcoholism. Doctors subjected patients who might be hypoglycemic to extensive testing, or referred them to specialists for even more costly evaluation.

David Nathan, associate professor of medicine at Harvard Medical School and director of the Diabetes Research Center at Massachusetts General Hospital, said hypoglycemia became a "wastebasket diagnosis" that physicians used to reassure patients who wanted a name for their unidentifiable and upsetting symptoms. Today medical experts agree that not only was hypoglycemia a fad label, but also that the old method of testing for low blood sugar was inaccurate and grossly inflated the number of cases.

The type of hypoglycemia once thought to be relatively common—characterized by headache, fatigue, tremulousness, sweating, palpitations, and an inability to concentrate several hours after a high-carbohydrate meal—is actually rare. Using more reliable testing methods, physicians at Massachusetts General's busy diabetic clinic, for example, see only a few patients a year who fit the criteria for this disorder, called *reactive hypoglycemia* because it occurs after eating. A more serious condition, called *fasting hypoglycemia,* occurs in response to going without food for too long. This type results from certain diseases that upset the body's ability to balance blood sugar.

These days the only people labeled as reactive hypoglycemic, the type that causes the most confusion, are those whose symptoms occur while their blood sugar is in the hypoglycemic zone—below 50 milligrams per deciliter (mg/dl)—and who improve immediately if they eat something.

Nevertheless, some clinical nutritionists believe that sudden fluctuations in blood sugar level can cause symptoms even in people whose blood glucose never dips below the 50 mg/dl threshold. Other experts disagree, saying that although these symptoms occur, sudden shifts in glucose levels are not to blame.

Testing the Limits

For years the standard test for patients suspected of having low blood sugar was the *oral glucose tolerance test* (OGTT), which requires drinking 75 to 100 grams of glucose solution and then having blood drawn every 30 minutes for 5 hours to determine the circulating sugar level.

Although a two-hour OGTT is a highly accurate method for diagnosing diabetes, researchers have doubted its value for confirming hypoglycemia since the late 1970s.

Reprinted from the November 1994 issue of the *Harvard Health Letter,*
© 1994, President and Fellows of Harvard College.

Some patients who test below 50 mg/dl have no symptoms of low blood sugar; others have symptoms, but their glucose levels are within the normal range (80-120 mg/dl).

In 1989, a study published in the *New England Journal of Medicine* determined that reactive hypoglycemia cannot be accurately diagnosed by administering an OGTT. Researchers from the University of Montreal showed that blood glucose levels are reliable only if blood samples are drawn while the patient is experiencing symptoms. This new approach to testing "put the nail in the coffin" for hypoglycemia's popularity as a diagnosis, turning up far fewer cases than physicians expected based on the OGTT or patients' reported discomfort after a meal, according to Lloyd Axelrod, senior physician and endocrinologist at Massachusetts General Hospital.

Today, patients with suspected hypoglycemia are usually told to eat the foods that typically provoke symptoms, and then to visit the doctor's office to have blood drawn and tested. Some physicians lend patients a *glucometer* (a device for measuring glucose) to use at home. Most of the time, blood sugar levels turn out to be in the normal range—not in the hypoglycemic zone.

Still, patients who have the classic symptoms are reacting to *something*, Dr. Nathan said, and it's important to find out what's going on in order to separate a real medical problem from a fad diagnosis.

Facing the Facts
The brain depends on glucose for energy. Since it can't produce its own supply, it needs to have a continuous influx of glucose from the bloodstream.

In healthy people, blood sugar is kept fairly constant by the interaction of several body mechanisms that act as checks and balances to bring the system back to equilibrium. Consuming carbohydrates, whether in the form of sugary sweets or starchy potatoes, normally stimulates the pancreas to produce insulin. The surge of insulin in turn causes the circulating glucose level to fall. As the concentration of blood sugar drops, insulin returns to normal levels.

There are several reasons why this feedback system can get out of whack. In people with fasting hypoglycemia, for example, liver disorders, insulin-producing tumors (insulinomas), or adrenal gland failure can throw off the normal glucose-insulin balance. Low blood sugar can also be brought on by some drugs, such as insulin or oral medications for diabetes mellitus, or as a result of stomach surgery.

The hypoglycemia-like symptoms that some people may experience after a high-carbohydrate meal don't actually reflect a disorder in the insulin-glucose balance, but are most likely related to dietary habits, stress, or emotional factors, according to nutritionist Connie Roberts, manager of nutritional consultation and wellness programs at Boston's Brigham and Women's Hospital. Swings in blood sugar levels, rather than demonstrably low glucose, are likely to cause apparent hypoglycemia, she said. Although anecdotal evidence suggests that some people may experience symptoms above the hypoglycemic threshold, studies have not been able to demonstrate these effects.

Getting Even
Those who experience symptoms even though their blood sugar level is normal may find that changes in diet can get them back on an even keel.

Skipping meals, giving in to a late afternoon craving for a chocolate bar, or indulging in caffeine or alcohol can put someone on a metabolic roller coaster, according to Roberts. Caffeine initially boosts energy by stimulating production of epinephrine, but it also increases the brain's use of glucose at the same time that it decreases cerebral blood flow, shortchanging the brain's usual energy supply. A study published in the *Annals of Internal Medicine* in 1993 showed that people who consumed caffeine could develop hypoglycemic symptoms even at blood sugar levels within the low-normal range. Excessive alcohol consumption can also cause these symptoms by interfering with the liver's ability to supply glucose to the body.

Often what helps is turning feast-or-famine people into grazers. Eating several small meals throughout the day guarantees a more even level of blood sugar over time, said nutritionist Roberts. She usually recommends that people eat a diet that is moderate in lean protein (chicken and fish), and that emphasizes complex carbohydrates, (breads and cereals), rather than simple sugars (candy, sweet rolls, or fruit juice).

Dietary change may be one solution to the fatigue, mood swings, headache, and neurologic symptoms that some people chalk up to hypoglycemia. In other cases, stress or emotional problems, such as anxiety or depression, could be to blame. Because these vague symptoms can be traced to many possible causes, it is usually wise to see a doctor to determine the root of these problems.

So consider skipping the donuts and coffee on the road and grabbing a bagel and a cup of herbal tea instead. That, plus taking a few minutes to breathe deeply and unkink those neck muscles, might just do the trick. ∎

Questions

1. What symptoms are associated with hypoglycemia?

2. How do physicians now test for hypoglycemia?

3. What types of conditions can disrupt the glucose-insulin balance?

Answers begin on page 161.

42 *Lactose intolerance is an inability to digest the lactose found in milk and other dairy products. In pronounced cases, a glass of milk, a slice of pizza, or a dish of ice cream produces bloating, stomach cramps, gas, or diarrhea. Lactose intolerance is caused by a deficiency of lactase, the digestive enzyme responsible for the hydrolysis of the disaccharide lactose to the monosaccharides glucose and galactose. Some experts estimate that as many as 50 million Americans suffer some degree of lactose intolerance, but a recent report says that number is a gross exaggeration.*

Marketers Milk Misconceptions on Lactose Intolerance

It would appear as though lactose intolerance is on the rise. Purchases of reduced-lactose milk as well as pills and drops that help the lactose-intolerant stomach dairy foods have been going up steadily during the last couple of years, with sales jumping to $117 million in early 1993, a 27 percent increase over the year before. Self-treatment for the problem has become so prevalent that *Rolling Stone* magazine has referred to it as "the hot disorder of the 90s."

But the current "spread" of the condition, which is an inability after early childhood to fully digest the sugar in milk—lactose—that can lead to gas, bloating, stomach cramps, and diarrhea, has nothing to do with any true epidemic. It stems more from marketing hype.

Advertisements for at least two products have suggested that 50 million or more Americans suffer stomach problems when they drink milk or eat ice cream or other dairy products, the major source of calcium in the U.S. diet. And people are buying into it. But that number is a gross exaggeration. Here's why.

Intolerance versus Maldigestion

One of the biggest misconceptions surrounding lactose intolerance concerns the meaning of the very word intolerance. Companies selling reduced-lactose milk and similar products use the term lactose intolerance to describe the condition of all people who cannot properly digest milk sugar. But what the companies call lactose intolerance, researchers in the field call lactose maldigestion. The difference is that lactose maldigestion, unlike intolerance, is not necessarily accompanied by unpleasant gastrointestinal symptoms.

In fact, only a small fraction of the estimated 50 million lactose maldigesters have an out-and-out intolerance—that is, stomach upset upon consumption of dairy products. In other words, while your body might not be able to fully "process" the milk sugar lactose even though it digests and absorbs the other nutrients in dairy foods, chances are you will not suffer gastrointestinal problems as a result of eating milk-based products.

Consider that when we are born, we produce an enzyme called lact*ase* that allows lactose to be digested in the small intestine. As babyhood is outgrown, however, many people begin to stop producing lactase. That means rather than breaking down, lactose can end up lingering in the gut, drawing in water and serving as food for gas-producing bacteria that "hang out" there. The potential (but not definite) result: bloating, flatulence, abdominal cramps, and diarrhea, which can occur anywhere from about 30 minutes to a couple of hours after dairy foods have been consumed.

Lactose maldigestion is actually one of the most common gastrointestinal disorders on a worldwide basis. The only individuals *not* very likely to have it are those of northern European ancestry (most Americans) along with those descended from peoples in a few pockets of the Mediterranean and the Near East as well as parts of Central Africa and the Indian subcontinent.

That means the majority of human beings, including tens of millions in the U.S.—most blacks of African descent, those of Asian ancestry, many people whose first language is Spanish, Native Americans, and Jews—are more likely to be lactose maldigesters than not.

But, as stated before, not by any means does everyone who maldigests lactose turn out to be lactose intolerant. One reason is that many maldigesters do not lose

Reprinted with permission, *Tufts University Diet & Nutrition Letter*. Subscription information: 1-800-274-7581.

all ability to produce the digestive enzyme lactase; they continue to make it in reduced amounts, which allows at least some dairy products to be properly digested. In addition, there are "friendly" bacteria in the gut that digest a portion of lactose even in the absence of lactase.

Then, too, there appears to be a "use it or lose it" aspect to handling lactose. That is, research suggests that people with lactose maldigestion can maintain a certain ability to digest milk by *not* avoiding dairy products altogether. The hypothesis is that the consistent consumption of milk-based foods allows the colony of friendly, lactose-digesting bacteria in the gut to stay "strong" as well as great in number at the same time that it causes the gas-producing bacteria to "falter."

That's what researchers at Meharry Medical College in Nashville, Tennessee, suspected when they gave increasing amounts of lactose to a group of lactose-maldigesting African Americans. Each one started out with a milk drink that contained just 5 grams of lactose per cup (a cup of milk normally contains more on the order of 12 grams of lactose, or three teaspoons' worth). If he or she did not suffer any unpleasant symptoms for two to four days, the lactose content of the milk beverage was increased by one gram for another few days.

The procedure kept being repeated until a lactose dose was reached that did provoke symptoms of intolerance. If the symptoms were extremely uncomfortable, a lower lactose dose was given, but it was gradually increased again until the maximum tolerated dose was determined.

After just six to 12 weeks, every single participant was able to tolerate at least the amount of lactose in a half cup of milk, and more than three out of four were able to tolerate at least a cup of regular milk at a time—even though they had started out not being able to drink a cup of milk without feeling uncomfortable. In other words, much of their "intolerance" was overcome.

The findings at Meharry underscore the experience of those who have participated throughout the world in supplementary feeding programs for children who were predominantly lactose maldigesters. In Ethiopia, for example, youngsters who suffered unpleasant symptoms during the first week of drinking a cup of milk every day tolerated their milk without a problem after that. And in Colombia, children who experienced symptoms when milk was first introduced at their school were able to tolerate it without incident six months later.

It Could Be Just a 'Gut Reaction'

Sometimes a person perceives he has a lactose intolerance not because of an inability to digest lactose but simply because of a *belief* that milk products disagree with him—a belief embedded in cultural norms.

The researchers at Meharry Medical College demonstrated this when they checked for lactose maldigestion in more than 150 African Americans who claimed that they could not drink a cup of milk without discomfort. Lab tests showed that in fact only 58 percent of them were true lactose maldigesters. What's more, when they further tested some of the maldigesters for symptoms by giving them both regular and low-lactose milk, and didn't tell them which was which, fully one in three reported unpleasant reactions no matter which beverage was drunk. In other words, perhaps just the *idea* of milk made some of them feel sick.

The notion that milk can be disagreeable may be particularly pronounced among certain ethnicities or cultures. If a person is of, say, African or Asian descent, and therefore is a member of a group in which lactose intolerance is prevalent, an unpleasant reaction to milk may be "learned" whether the individual is truly lactose intolerant or not.

Getting Tested for Lactose Intolerance

One increasingly popular way to try to determine whether you are lactose intolerant is to use a free "test kit" offered by Johnson & Johnson, which, it so happens, markets lactose-reduced milk under the name Lactaid was well as pills that contain the enzyme necessary for digesting milk sugar. Called The Lactaid Challenge, the kit consists of three Lactaid tablets and instructions to eat a normal breakfast plus a 12-ounce glass of milk. It then instructs the consumer to eat the same breakfast the next morning, but only after having taken the three tablets.

"If you are lactose intolerant," the kit says, "you'll probably notice less, or even none, of the discomforts you experienced on Day 1. If the discomforts continue, you may need to increase the amount of Lactaid you take."

To "help" others "diagnose" themselves, the company may include with the kit a form on which to fill in the names and addresses of two friends and then mail back so that your pals, too, can be sent a free "test."

The problem, of course, is that the test has the power of suggestion built into it. "If milk never bothered you before, someone's advice to take a pill to overcome potential problems digesting it could certainly make you believe it bothers you now," says Robert Russell, MD, an associate director of the Human Nutrition Research Center at Tufts who specializes in digestive disorders. "That's especially true if you're not in the habit of drinking 12 ounces of milk at a time, which is the case for most Americans," Dr. Russell points out.

So what *is* the best way to test for lactose intolerance? Some say a diagnosis should be as simple as the old vaudeville joke:

Patient: Doctor, it hurts when I do that.
Doctor: Don't do that.

In other words, if you suspect that dairy foods disagree with you, avoid them. But that approach is flawed. One reason is that lactose intolerance is sometimes nothing more than a temporary condition. It can occur as a result of such illnesses as inflammatory bowel disease, a parasitic infection, or even stomach flu, all of which can induce a short-lived deficiency of the enzyme lactase that is needed for the digestion of lactose. Sometimes, taking certain medications, such as various antibiotics, can also bring on temporary lactose intolerance by causing a short-term lactase deficiency. Whatever the mechanism, the fact that lactose intolerance can be a passing condition in some individuals makes the advice to simply "avoid" milk products a much more limited solution than is often necessary.

It's especially important not to automatically declare dairy foods off limits for children, who in our culture depend on milk-based products for a host of essential nutrients, including not only calcium but also protein, riboflavin, vitamin D, and phosphorus, among others. If your child does suffer from ongoing digestive problems, ask your pediatrician for advice. In other words, don't ban milk products on a hunch.

In some cases, of course, the hunch will be correct, and a physician, whether in dealing with a child or an adult, will able to determine that the patient does have lactose intolerance without a lab work-up. After all, if a person consistently suffers gas or bloating even after eating just a piece of milk chocolate or a dish to which a bit of milk has been added, a diagnosis of lactose intolerance can be made without extensive testing. But in cases where the doctor is uncertain, a diagnostic test may be in order. One such test involves drinking a large dose of the milk sugar lactose dissolved in water on an empty stomach and then measuring the rise in blood sugar. The higher the rise in the level of sugar in the blood, the more lactose that has been digested and therefore the less apt the person is to suffer from true lactose intolerance.

An even more sensitive diagnostic procedure is known as the breath-hydrogen test, which runs about $200. First, the physician measures the hydrogen concentration of the breath on an empty stomach—before any lactose is consumed. Then, he or she gives the patient some lactose to drink in a watery solution. If the patient is unable to digest it, the hydrogen concentration of the breath will increase significantly. The reason is that the "bad" bacteria, which feed off lactose when it is not digested, break the lactose molecules apart in such a way that hydrogen is produced, and then makes its way to the respiratory system.

Depending on how high the level of hydrogen rises, the doctor can get a handle on whether a person's symptoms of discomfort upon eating dairy foods correlates to an impaired ability to digest lactose.

Having Your Ice Cream and Eating It Too

The good thing about lactose intolerance is that it's a very "forgiving" condition. That is, there's a lot that even people who have been diagnosed with the problem can do to make the consumption of nutrient-rich milk and other dairy products pain-free.

Drink milk more often but in smaller amounts. That allows whatever lactase is present to do its job of digesting the lactose before it starts causing problems. It also gives the "good" bacteria in the gut more of a chance to handle the lactose load present at any given time.

Have milk and other dairy foods with meals or snacks. That slows the rate at which the stomach empties and, again, gives the friendly bacteria smaller doses of lactose to digest at one time. (The crust on cheese pizza, the eggs and other ingredients in ice cream, and the ingredients in a cheese-topped casserole dish have the same digestion-slowing effect.)

Gradually increase consumption of milk-based foods. As studies suggest, that appears to acclimate the digestive system to lactose over time.

Eat aged or hard cheeses, such as Swiss and cheddar. Most of the lactose is removed when these cheeses are made. Also, eat yogurt with active cultures (which is another way of saying live bacteria). The bacteria in many brands of yogurt provide their own lactose-digesting enzymes and continue to digest the yogurt's lactose even further once inside your digestive tract. Be sure to let the yogurt sit at room temperature for a half hour before you eat it. The bacteria are dormant, or inactive, at refrigerator temperatures.

Note: Some yogurt labels boast that the product contains acidophilus cultures, but several strains of those bacteria are not responsible for digesting lactose; other microorganisms in yogurt do the trick. The cultures in acidophilus milk do not digest lactose either, so it is not a ready alternative to regular milk.

As a last resort, consider going through the added expense of buying reduced-lactose milk and cheese, or droplets that you can add to milk to digest most of the lactose for you before you drink it, or enzyme-containing capsules or pills you can take before eating dairy foods.

Reduced-lactose milk, sold under such names as Lactaid and Dairy Ease, generally contains only about a third of the lactose of regular milk (although some are lactose-free) but is almost twice as expensive—$1.39 in Boston-area stores. Adding enzyme droplets to milk and letting it stand for 24 hours eliminates about two thirds of the lactose (even more lactose is broken down if you let the milk stand for 48 to 72 hours). It's also a less expensive alternative. It adds about 20 to 60 cents to the price of each quart of milk, depending on how many droplets you use. Both reduced-lactose milk and milk treated with droplets taste sweeter than regular milk. That's because when lactose is broken down, it turns into sweeter forms of sugar.

As for enzyme tablets taken just before having dairy foods, it takes about 55 cents worth to break down most of the lactose in a glass of milk. But they're good for eating out or at a friend's house—or if you feel like buying an ice cream cone from a street vendor on the spur of the moment.

It should be noted that among those afflicted with lactose intolerance, there is a very small subset of individuals for whom the special capsules and droplets may not help because of an acute sensitivity to even the smallest amounts of milk sugar. Such people feel uncomfortable eating foods with just tiny amounts of lactose, including many breads and other baked goods, processed breakfast cereals, instant potatoes and soups, certain brands of margarine and luncheon meats, salad dressings, nondairy creamers and whipped toppings, and mixes for pancakes, biscuits, and cookies. These individuals (some of whom are born with an extremely rare congenital inability to produce the lactose-digesting enzyme lactase) need to carefully read the ingredients lists on food labels, looking for not only the words milk or lactose but also whey, curds, casein (as well as sodium caseinate), lactalbumin, milk by-products, dry milk solids, and nonfat dry milk powder. All of these contain lactose.

So do about 20 percent of all prescription drugs (including many types of birth control pills, for example) and 6 percent of over-the-counter medicines, like certain tablets for stomach acid and gas. That's why the severely lactose intolerant also need to read labels on products they buy from a drugstore as well as speak with their physicians and pharmacists about their condition.

Is It an Intolerance or an Allergy?

Most of the world's population maldigests lactose to one degree or another. But some individuals have trouble with milk-based products not because they lack a digestive enzyme but because they are *allergic* to milk or, more specifically, to milk protein. In reaction to dairy products, their immune systems cause such allergic symptoms as rashes, hives, wheezing, redness, stuffy nose, and runny eyes.

Unlike lactose intolerance, which tends to start in toddlerhood, most milk allergies occur in infancy and are *outgrown* after toddlerhood. And milk allergy is rare, affecting only about two to three out of 100 children under age three. By the time adulthood is reached, only about one out of 200 persons remains allergic to milk and other dairy foods.

To diagnose a milk allergy, a physician relies on a different set of tests than those used to detect lactose maldigestion. People who do turn out to be allergic cannot acclimate themselves to dairy products the way lactose maldigesters can; they must generally avoid them. ■

Questions

1. What is the difference between lactose maldigestion and lactose intolerance?

2. What conditions can cause temporary lactose intolerance?

3. How does measuring blood sugar after drinking a lactose solution indicate whether a person is lactose intolerant?

Answers begin on page 161.

43 *The science of nutrition involves studies that have a great deal of practical significance. For example, nutritionists might try to determine the proper components of a sound diet or the best way to maintain proper body weight. Popular magazines and newspapers sometimes add confusion to this area of science with incomplete or inaccurate articles. The attention-grabbing headline of a recent* New York Times *article probably reinforced the false belief of many people who consider foods rich in carbohydrates to be fattening.*

Nutrition: Pasta Is Not Poison

Kristine Napier

It was a story that set Italy on its heel. Bearing the title, "So it May Be True After All: Eating Pasta Makes You Fat," a February 8th *New York Times* article shocked Americans with the news that a low-fat, high-carbohydrate diet causes some people to gain weight. According to the *Times,* eating pasta and other carbohydrate-rich foods causes one in four Americans to produce excessive amounts of insulin. In turn, declared the *Times,* insulin transforms even small amounts of carbohydrates into body fat at a mysteriously rapid rate. *Hyperinsulinemia* (elevated insulin level) can contribute to the development of heart disease and hypertension.

Italians called the article an "anti-spaghetti conspiracy," and one Italian nutritionist said that it reminded him of a news story published in a reputable French paper that said that drinking Coca-Cola makes women give birth to daughters.

Some scientists say that the *New York Times* article was a case of strange bedfellows—two peripherally related stories woven into one inaccurate and confusing whole. "The article mixed up apples and oranges," said Gerald Reaven, a professor of medicine at Stanford University Medical School. "The story was probably supposed to be about how Americans are eating lower fat diets but aren't losing weight. Insulin resistance is a different issue. While some overweight people subsequently become insulin resistant, the two are sometimes totally unrelated."

George Blackburn, associate professor of surgery at Harvard and director of nutrition services at Boston's Deaconess Hospital, agrees. "If the two are related it's that insulin resistance is sometimes the consequence of obesity, not the cause of it."

The good news is that pasta and other carbohydrate-rich foods will not make people fat—unless they eat too much of them.

Between 1978 and 1990, Americans cut their average intake of total dietary fat from 36% to 34% of calories—which works out to one less teaspoon of butter or margarine daily for someone consuming 2,000 calories. This continues a trend that began in the 1960s when fat accounted for more than 40% of calories in a typical American diet. According to the National Heart, Lung, and Blood Institute, this shift in diet helps explain why average cholesterol levels have also changed for the better.

Low-Fat Does Not Equal Low-Calorie

At the same time, national surveys show that the daily calorie intake for Americans increased from an average of 1,969 in 1978 to 2,200 in 1990. "While people who want to drop weight have decreased dietary fat, they've also become volume eaters, giving in to a second or even third bagel because they're using no-fat cream cheese," said registered dietitian Cathy Nonas, director of the Theodore van Itallie Center for Weight Control at St. Luke's Roosevelt Hospital Center in Manhattan.

"Too many Americans have been lulled into a false sense of security with the explosion of fat-free alternatives, and they don't realize that low-fat doesn't mean low-calorie." As a result, more people are now officially considered obese: they weigh at least 20% more than experts say they should. In only a decade, the percentage of obese adults has gone from 25.4% to 33.4%. "It's a case of too much en-

Reprinted from the July 1995 issue of the *Harvard Health Letter,* © 1995, President and Fellows of Harvard College.

ergy in as food and too little energy out as exercise," said Stanford's Dr. Reaven.

Indeed, studies support this explanation. One survey showed that the percentage of adults who exercise or play sports on a regular basis decreased steadily between 1985 and 1990; another revealed that, in 1991, more than 58% of Americans reported that they seldom or never engaged in leisure-time physical activity. The link between sedentary behavior and obesity is well known. One recent study showed that body mass index (a measure of obesity) increased along with time spent watching television. Not surprisingly, people who watched TV the most exercised the least and had the highest body mass index.

"You're no more likely than anyone else to gain weight if you're insulin resistant. You are, however, more likely to suffer potentially dangerous side-effects," said Dr. Reaven. Like anyone else, insulin resistant people only gain weight if they eat too much and exercise too little.

What Is Insulin Resistance?
Some foods are more efficient sources of energy than others: 100% of carbohydrate, 58% of protein, and 10% of fat is turned into glucose, which fuels the body. But glucose would be useless without insulin, which shuttles it into cells. In the estimated one in four Americans who are insulin resistant, cells won't let glucose in unless there's an abnormally large quantity of insulin on hand to open the door. In other words, extra effort is needed to overcome the resistance. "It's like the difference between pushing a car uphill or on a flat road," said Harvard's Dr. Blackburn.

As a result, insulin resistant people frequently have higher than normal amounts of insulin circulating in their blood. In some of them this hyperinsulinemia progresses to the development of adult-onset diabetes. More often it both raises triglycerides and lowers helpful HDL cholesterol, a sinister combination that significantly increases the risk for artery-clogging heart disease and high blood pressure. High insulin levels also raise cardiovascular risk by making it easier for clots to form and may damage the lining of blood vessels, both of which make blockages more likely.

Both nature and nurture play a part in insulin resistance: some people have a genetic predisposition to it that kicks in even if they are lean; others develop the condition as a consequence of being overweight and sedentary. The good news is that the latter group can eradicate the problem by losing weight and exercising.

Just Say No
For most overweight people, "pasta and other carbohydrates with vegetable toppings continue to be a solution to obesity, not the cause," said Dr. Blackburn. "But portion control is crucial." How many people know, for example, that some bagels from speciality shops are the equivalent of four slices of bread or four starch servings? Or that when dietary guidelines speak of a serving of pasta they mean half a cup? (For a rude awakening, measure that out onto your dinner plate!)

Nutritionists warn that pudgy people on the go may actually be favoring highly processed carbohydrates over other, more nutritious foods. "In their hurry, many grab the readily available starch items such as bagels, pretzels, and rice cakes, forgoing the more complex carbohydrates such as fruits, vegetables, and grains," said Nonas, the nutritionist. Although current dietary guidelines recommend eating five to nine servings of fruits and vegetables daily, a National Cancer Institute survey showed that 42% of Americans eat three or fewer servings and only 23% consume five or more.

"If Americans would become volume produce eaters instead of volume starch eaters as they decreased fat intake, we wouldn't see bloating waistlines," Nonas said. Two cups of Romaine lettuce tossed with tomato wedges, carrots, red bell peppers, and fresh strawberries has 144 calories, compared with 600 calories in two bagels from the corner bakery. Produce is not only a low-calorie way of filling up, but also a much better source of fiber, vitamins, and a plethora of phytochemicals. (See "Cancer Fighting Foods," *Harvard Health Letter,* April 1995.) "High-fiber foods also squelch appetite for a much longer period" than typical high-carb snacks, noted registered dietitian Jill Stovsky of the Cleveland Clinic Foundation.

Nonas recommends the following: "Half of your dinner plate should consist of steamed vegetables, one-quarter of it should be starch, such as rice or pasta, and one-quarter of it lean meat or legumes. Vegetables should become the focus of the dinner plate, not an afterthought." About 50-60% of the day's total calories should come from carbohydrates in the form of fruits, vegetables, and whole grains.

As for overweight people who are insulin resistant, weight loss should remain the primary goal. Normal-weight individuals with the condition should talk to their doctor about the optimal diet for them. It may well involve eating slightly more fat, mainly in the form of olive and canola oils and other unsaturated fats.

Despite the attention-grabbing headlines about pasta and portliness, nutritionists say that there's nothing magical about how to lose weight. "If you eat more calories than your body expends, you'll gain weight. If you eat fewer calories than you burn off, you'll lose weight. It doesn't matter where those calories have come from," said Nonas. ■

Questions

1. Do pasta and other carbohydrate-rich foods make people fat?

2. What is the clinical definition of *obesity?*

3. What is insulin resistance, and what causes it?

Answers begin on page 161.

44 *Scientists have moved a step closer to developing a drug that could treat obesity in humans. Three separate research papers appearing in the journal* Science *describe advances made since a team in December 1994 identified the ob gene believed to make mice obese. Injections of a protein hormone derived from the ob gene helped overweight mice lose 20–30 percent of their weight in 30 days. Even lean mice with no genetic problems lost some weight when given the treatment. Researchers compared the hormone, which appears to regulate the storage of fat, to insulin, which controls sugar levels. Weight loss is a multibillion-dollar industry, and Amgen, the California company that developed the protein, has filed for a patent. Amgen calls the hormone* leptin *and hopes to begin safety testing on humans next year.*

The New Skinny on Fat

Traci Watson

A protein that makes mice shed pounds might ultimately help people, too.

If Jenny Craig and Richard Simmons ever have job-related nightmares, the monster in their dreams must resemble the "ob mouse." The ob stands for obese, and, true to its name, the ob mouse is immense—up to three times as big as a regular mouse. Bred to be overweight, it is cursed with both a slow metabolism and a teenage boy's appetite.

But dietary salvation has now arrived for the ob mouse. In last week's issue of *Science,* three separate research teams announced the discovery of a slimming potion for the hapless mouse. The magic chemical is a protein called OB, and daily injections over several weeks were able to transform a naturally grotesque ob mouse into something approaching the mouse ideal.

The announcement caused a sensation in both scientific and financial circles and triggered a flurry of headlines about miracle diet pills. "These are very exciting results," says Claude Bouchard, an obesity researcher at Laval University in Quebec. "They're going to keep us talking for a while."

But most researchers cautioned that it's too early to know for sure if OB could really contribute to a human obesity treatment. They caution that the protein will have to undergo years of testing before humans can take it—and it cannot be made into something as simple and easy to take as a pill. Even then, OB may have side effects, and some scientists say it's never likely to provide a post-holiday or pre-college-reunion fat cure.

There are several things that make obesity researchers hopeful they are on the right track with OB, however. For one, the protein appears to work by both diminishing appetite and speeding up metabolism. Scientists speculate that the protein's natural role in the body is to tell the brain when the body's fat stores are growing; the brain, in turn, commands the body to lower food intake.

The ob mouse has been lolling around laboratories since the 1950s, when it first appeared by chance in mouse breeding colonies. But not until the 1960s and '70s did a series of grotesque experiments show that a biological compound could cause an ob mouse to lose weight. Douglas Coleman, a scientist at the Jackson Laboratory, the world-famous mouse-breeding facility in Bar Harbor, Maine, surgically coupled an ob mouse and a normal mouse to create a rodent version of Siamese twins. The fat half of the pair soon became skinny. Coleman realized that some kind of blood-borne factor must regulate weight. But the scientific techniques of Coleman's day were too blunt to find the flab-controlling substance.

In 1987, with molecular biology techniques much improved, scientists from the Howard Hughes Medical Institute at Rockefeller University began a concerted effort to find the gene responsible for the ob mouse's plight. It took them nearly eight years, but they finally tracked it down, and announced their discovery last December. The chunky mice, it turned out, all had a mutation in one gene, also designated ob. In these mutant mice, the

Copyright, August 7, 1995, *U.S. News & World Report.*

ob gene malfunctions and fails to produce the OB protein. These mice grow large in part because their brains never get the message that the body already has lots of fat. As a result, the brain never tells the mouse to stop eating.

The Clincher
Having found the ob gene and the OB protein it produces in normal mice, the Hughes-Rockefeller scientists and teams from the drug company Hoffman-La Roche and the biotech firm Amgen performed the obvious experiment to clinch the connection: They gave injections of OB protein to the naturally fat ob mice. They they sat back and watched the ounces melt away. Not only were the mice eating less but their metabolism was also speeding up. They lost weight; their fat stores shrank; their blood sugar, which bordered on diabetic levels, fell.

Why all this happened is still not precisely understood. The OB protein appears to have many roles in the body, and scientists aren't even sure which organs it targets. One effect does seem sure: OB shots finally tell the ob mouse's brain to shut down feeding. "This animal is fat because he thinks he's starving," says Jeffrey Friedman, leader of the Hughes-Rockefeller team. "When you give him the protein, the animal gets thinner because he thinks he's getting fatter."

Even more promising to those hoping for a practical fat therapy was OB's effect on so-called diet-induced obese mice. These are naturally svelte mice bred to have a fat tooth—and to gain weight quickly on a high-fat diet. So the Hoffman-La Roche team fed their mice what team leader L. Arthur Campfield compares to "chocolate-chip cookie dough without the chocolate chips." Faced at every meal with the mouse equivalent of cheesecake, the mice overate and grew quite large. But when given shots of OB protein, the mice slimmed down in a hurry.

Also cheering was the finding that even lean mice unfattened by genes or diet lost weight on the OB diet plan. And at least some preliminary hints that what works for mice might work for men were further cause for optimism. The Hughes-Rockefeller team had already found last December that humans carry the ob gene. When the team gave *human* OB protein to ob mice, the mice still lost weight. So the human form of OB must work in a similar way to the mouse form.

A mutation in the human ob gene, researchers speculate, may be at the root of at least some cases of human obesity. These are the people that most scientists agree could almost certainly be helped by treatment with OB. But these would probably be people who are morbidly obese—weighing 300 to 400 pounds, a condition that is extremely rare.

People who are mildly obese, on the other hand, are hardly rare at all: A recent government survey found that one third of Americans are overweight. The reasons for that statistic are both environmental and genetic. Americans live in the land of milkshakes and honey-dipped doughnuts—and physical inactivity. Our genes prompt us to gain weight on this regimen. Thirty to 40 percent of obesity is due to genetics, scientists estimate. Indeed, a recent editorial in the *New England Journal of Medicine* dismissed as a "folk belief" the idea that overeating alone causes obesity.

Roadblocks
Whether OB can help people whose obesity is due to both genetics and lifestyle is up in the air. On the one hand, the weight loss that the diet-induced obese mice experienced on OB was promising. But fatness takes a thousand forms; there are a half-dozen known genes causing obesity in mice. And at least some of these forms of genetic obesity would definitely not be helped by the OB treatment.

The db (diabetes) mouse is a case in point. These mice are every bit as fat as ob mice, but when injected with the OB protein, the db mice stayed fat. Researchers suspect that these mice are OB "insensitive"—that is, their cells lack the ability to notice whether the OB protein is present. So even though the mice make their own OB, their bodies do not heed its effects, and extra protein doesn't help. Some overweight humans may very well have a similar problem, rendering them impervious to the powers of surplus OB.

Many other hurdles loom. At present, OB can be taken only by daily injection. Amgen, which is pursuing OB as a drug, hopes to come up with another way patients could take the drug—perhaps an implant or pump. An oral form is not possible, however, since the OB protein molecule is broken down by the digestive process before it can be absorbed through the gut. Worse, OB's effects last only as long as the drug is taken religiously: One research team took their mice off OB protein for only two days—and watched the mice's weight shoot up.

Side effects could also prevent OB from entering the mass market. The protein seems likely to be a hormone, and hormones, such as insulin, often have wide-ranging effects on the body. If this is the case for OB, long-term use of the drug could spring unpleasant surprises. Mice given too much OB protein waste away and die. Tests for side effects in humans should start in 1996, followed by tests of OB's effectiveness. If all goes well, OB could enter the market in five to seven years.

Luckily, if OB doesn't pan out, scientists have lots of other ideas, many of which involve pills. It may be possible, for example, to find a chemical that boosts the body's own production of OB. Researchers are also investigating ways to turn up the body's sensitivity to OB, which would benefit db mice and any humans with a similar genetic defect. But these drugs would require at least an extra five years to develop.

Other therapeutic avenues may be opened as more obesity genes are discovered. The latest fat gene was announced last week by a separate group of researchers. Officials at Millennium Pharmaceuticals of Cambridge, Mass., said their scientists had found the mouse and human versions of a gene dubbed "tub." Tubby mice, as they're known, never get quite as Rubenesque as ob mice. And while the ob mouse is fat from Day One, the tubby mouse fattens slowly after reaching puberty. That's how most people gain weight, too, so scientists are anxious to find out more about how tub acts.

Doctors worry, of course, about misuse of fat-melting miracle proteins. But even the worriers think the benefits outweigh the hazards because overweight people who slim down reduce their risk of diabetes and heart disease. And the ob mice saw another benefit, too, becoming livelier and more active—yet without losing their sweet, gentle natures. No human dieter could ask for more. ■

Questions

1. How is the OB protein believed to function?

2. What is the ob gene?

3. When can studies of the hormone's effects on humans begin?

Answers begin on page 161.

45 *Cholesterol, the most abundant steroid in the body, has received much attention because of a correlation between cholesterol levels in the blood and the disease atherosclerosis, or hardening of the arteries. The problems created by the buildup of cholesterol-containing plaque on artery walls are responsible for nearly two of every five deaths in the United States each year. Although our knowledge of the role played by cholesterol in atherosclerosis is still incomplete, the latest studies clearly show that cutting cholesterol can save lives.*

Cutting Cholesterol: More Vital Than Ever

Public-health organizations have long urged Americans to protect their hearts by lowering their blood-cholesterol levels. But some researchers have had strong reservations about that advice. While reducing cholesterol cut the death rate from coronary heart disease, no convincing study had ever shown that it reduced the overall death rate. In fact, several cholesterol-lowering studies found a perplexing *increase* in mortality from noncoronary causes, including cancer, suicide, and violence. Even the coronary benefits were apparently limited: Cutting cholesterol seemed merely to buy time while atherosclerosis—clogging of the arteries caused partly by high cholesterol—continued to worsen.

But recent studies, including a new, landmark trial, have finally shown that cutting cholesterol can indeed reverse atherosclerosis and save lives.

Lower Cholesterol, Longer Life

In November, Scandinavian researchers reported the momentous results of a multinational study of some 4500 people, ages 35 to 70. The volunteers all had high cholesterol levels and coronary disease. The researchers randomly assigned them to receive either simvastatin (*Zocor*), a powerful cholesterol-lowering drug, or a placebo. After five years, the drug group had more than 40 percent fewer fatal heart attacks than the placebo group—a sharper, faster reduction in coronary mortality than any major study had shown. Equally important, there was no increase in deaths from noncoronary causes. Overall, only 182 people died in the drug group, compared with 256 in the placebo group—a decrease of nearly 30 percent.

Because that study was meticulously designed, and because the increases in noncoronary mortality found in previous studies were small and biologically implausible, it seems likely that those increases were simply statistical flukes, and that reducing cholesterol can indeed reduce the overall death rate—at least in men. The new study included only about 800 women, too few to determine whether reducing cholesterol can save lives in females, too. However, the study did provide the first solid evidence that cholesterol reduction lowers women's risk of a fatal heart attack, providing strong hope that studies including more women will eventually demonstrate a drop in total mortality, too.

Rolling Back the Blockage

Until recently, no one knew whether reducing cholesterol helped prevent heart attacks merely by slowing the buildup of plaque deposits in the coronary arteries or by actually halting or even reversing that buildup. In August, researchers at the University of California at Berkeley analyzed the combined results of 10 clinical trials, involving a total of more than 2000 people, that had tried to answer that question.

In nine of the trials, one group took medications, changed their health habits, or both, in an aggressive attempt to improve their cholesterol levels. That meant reducing LDL (low-density-lipoprotein) cholesterol, the kind that tends to clog the arteries; raising HDL (high-density-lipoprotein) cholesterol, the kind that helps clear the arteries; or doing both. (The tenth trial tested an experimental operation on the intestines that reduces cholesterol levels.) The control groups all received standard, less aggressive treatment.

"Cutting Cholesterol: More Vital Than Ever." Copyright 1995 by Consumers Union of U.S., Inc., Yonkers, NY 10703-1057. Reprinted by permission from *Consumer Reports on Health*, February 1995.

After an average of about three years, plaque deposits in the coronary arteries had increased in only 29 percent of the aggressively treated volunteers, compared with 53 percent of the control group. And it actually shrank in 31 percent of the aggressively treated people, versus just 11 percent of the controls. The reduction in plaque size was quite small—only about 1 to 2 percent, on average. But the treatment had a big impact on coronary risk: In seven of the eight trials that evaluated risk, the group that received aggressive treatment had anywhere from 25 to 89 percent fewer coronary problems than the control group had.

To help explain that striking benefit, researchers from several of the trials reanalyzed the results and performed further tests. They concluded that the improved cholesterol levels—and possibly other benefits of the nondrug measures—did more than just affect the size of plaque deposits. The plaque became smoother and more stable, reducing the chance that one of the deposits would break open, sparking a chain of events that could trigger a heart attack. In addition, the cells lining the coronary arteries released fewer clot-promoting proteins. And the arteries themselves became more flexible, boosting the amount of blood they can deliver.

Healthy Habits, Healthy Arteries

In most of the atherosclerosis trials, the volunteers took drugs to lower their cholesterol levels. But in three of the trials, they reversed clogging without drugs, just by changing their health habits.

The most widely publicized of those "lifestyle" trials was led by researcher and author Dean Ornish, M.D., of the University of California at San Francisco. Ornish's volunteers ate an extremely low-fat vegetarian diet, which derived less than 10 percent of calories from fat and hardly any calories from saturated fat. (In contrast, the average American gets 34 percent of calories from total fat and 12 percent from saturated fat). The volunteers also stopped smoking, exercised regularly, and attended lengthy motivational sessions. At a recent conference of the American Heart Association, Ornish reported that after four years those rigorous steps had reduced levels of LDL cholesterol by a full 30 percent and shrunk almost three-fourths of the plaque deposits; in the control group, nearly all the deposits got worse.

Such a demanding program probably maximizes the chance of success. But other trials have shown that it's possible to reverse atherosclerosis without resorting to such drastic steps. In one study, German researchers found that frequent exercise plus a somewhat laxer diet than Ornish's shrank plaque deposits in one-third of the volunteers. In a British study, scientists achieved even better results using no special exercise and an even laxer diet, which allowed 27 percent of calories from total fat and less than 10 percent from saturated fat. That finding may overstate the potential effectiveness of such a moderately lean diet for Americans, who already eat less fat than the British do. Still, the diet contained virtually the same amount of fat as the one that's recommended by the U.S. Government for all healthy adults.

Who Needs to Cut Cholesterol?

The evidence that reducing cholesterol can help open the arteries, prevent heart attacks, and save lives makes that step more important than ever for men and probably for women, too. It's particularly important for coronary patients. They should aim at least for an LDL level below 100 mg/dl.

Virtually all researchers agree that middle-aged people who don't have coronary disease should also try to reduce elevated LDL levels. (Note that the benefits of cutting cholesterol in healthy people wane by age 65 to 75 and may disappear entirely by age 80.) However, researchers disagree on whether middle-aged people without coronary disease should aim for a maximum LDL of 160 mg/dl or 130 mg/dl. In view of the recent findings, Consumers Union's medical consultants urge healthy people to try at least to achieve the lower goal, especially if they have two or more coronary risk factors. In addition, CU's consultants advise all middle-aged people, as well as older coronary patients to try to raise their "good" HDL if it's lower than 35 mg/dl.

Drug therapy is the simplest most effective way to improve cholesterol levels—and the only proven way to reduce total mortality. But it's usually expensive. People with only moderately elevated cholesterol levels should generally try nondrug measures first. (Coronary patients with moderately high cholesterol should turn to drugs sooner than other people—sometimes without waiting at all.) Nondrug measures include losing weight to reduce LDL and boost HDL; minimizing consumption of animal fats, the main source of saturated fat, to lower LDL; and exercising and stopping smoking to raise HDL. ■

Questions

1. What were the findings of the Scandinavian study of 4,500 people?

2. In addition to affecting the size of plaque deposits, what do researchers believe decreasing cholesterol does to plaques?

3. What nondrug measures can be used to decrease cholesterol levels?

Answers begin on page 161.

46 *The human body is 42–70% water (by weight), the percentage depending on a number of factors, including age, sex, and especially the amount of body fat. The chemical reactions of the body that are necessary for the maintenance of life take place in the watery medium of the body fluids. A loss of too much water from the body might adversely influence those reactions and, in extreme cases, create life-threatening situations. Adequate fluid intake is essential, as this article for sports enthusiasts emphasizes.*

Water: The Ultimate Nutrient

Nancy Clark

Amid today's sports-drink craze, some thirsty athletes have forgotten about the basic, old-fashioned fluid replacer: water. Although generations of athletes have successfully quenched their thirst with H_2O, high-tech beverages now have a strong presence.

But plain old water is worth remembering. Here is a reminder of just how important water is in today's sports diet, and some guidelines on how much you need.

What to Know about H_2O

Water is a basic nutrient. It is essential for you to live and for your body to function properly. You can survive for weeks without food, but only for a few days without water.

Water makes up about 60% to 70% of your body weight. Muscle tissue is 70% to 75% water. Fat, in comparison, contains only about 10% to 15% water.[1] Water has very important functions in your body:

In *saliva* and *stomach secretions* it helps digest food.

In *body fluids* it helps lubricate the joints and cushion organs and tissues.

In *blood* it transports carbohydrates, fats, proteins, hormones that regulate metabolism (body processes), and oxygen to the working muscles. It also carries away waste products such as carbon dioxide, ammonia, and lactic acid.

In *urine* it carries waste products out of your body. Exercise increases the production of wastes. Dark urine carries a lot of wastes.

In *sweat* it removes the body heat you generate during exercise. Water helps regulate body temperature by absorbing the heat from the muscles and transporting it to the skin's surface. Loss of a pound of sweat (about a pint) equals about 250 calories of heat lost.

As you can see, a severe lack of water (dehydration) can play havoc with your body's ability to function at its best. During the first few hours of water deprivation, water is lost primarily from blood volume. Because about 90% of your blood should be water, one danger of getting dehydrated during exercise is it will take longer for nutrients to be transported to and from your muscles. Your sports performance will suffer.

With continued water deficit, your cells lose water, resulting in cellular dehydration. When cells become overheated, they experience dramatic changes that can impair how they work.[1] Water loss of 9% to 12% of your total body weight can be fatal. No wonder you want to prevent dehydration!

Are You Thirsty?

Thirst is your body's way of telling you it wants or needs liquids. Under normal resting conditions, thirst does an adequate job of helping you maintain water balance. If your body fluids become abnormally concentrated because you're lacking water, your brain receives a signal that makes you feel thirsty. But this happens *after* you are already a little dehydrated.

In some circumstances, the thirst mechanism isn't reliable. For example, among athletes, thirst can be blunted by exercise and overridden by the mind. That's why extremely active athletes should drink more than required to satisfy their thirst. The thirst mechanism in young children and older adults may also lack the sensitivity needed to match their fluid needs. They may not feel thirsty even though their bodies need fluids.

How Much Is Enough?

Because water is fundamental for survival, you need to drink enough water every day to replace the amount you lose through

---

Nancy Clark, "Water: The Ultimate Nutrient," *The Physician and Sportsmedicine,* **May 1995. Copyright 1995, McGraw-Hill, Inc. Reproduced with permission of McGraw-Hill, Inc.**

urine, sweat, feces, and the air you breathe. Even if you don't exercise, you lose about 12 ounces of water per day with simple breathing and another 24 ounces through the skin (perspiration that you don't even notice). Add strenuous exercise, and you could lose up to 4 pounds of water (2 quarts) per hour.

How much water you need per day is determined by how many calories you burn. Your body needs about 1 mL of water for every calorie that you expend.[2] This means if you are active and burn 4,000 to 5,000 calories per day, you will need 4,000 to 5,000 mL (4 to 5 L or roughly 4 to 5 quarts) of water or fluids every day. If you are an inactive person, you may burn 1,500 calories and require only 1.5 L of water.

A simpler way to determine water requirements is to weigh yourself before and after you exercise. If you have lost 1 pound of body weight, you need to replace it with at least 16 ounces (1 cup) of water. Or better yet, plan to drink at least 16 ounces of water *during* your next workout to *prevent* dehydration.

An even easier way to find out if you are getting enough water is to simply monitor your urine. If you urinate a significant amount regularly throughout the day, and if your urine is clear colored, you are drinking enough. For some athletes this may mean drinking 12 to 16 glasses of water. Others will need to drink far less. Urinating about every 2 to 4 hours is fine. If you have to visit the bathroom every hour, you are needlessly drinking more than required.

Beyond the Basics

For some people, water may seem boring. Research by the makers of sports drinks has shown that active people tend to drink more fluid if it tastes good. That is, given the choice between a tasty sports drink and plain water, the sports drink will disappear faster. A sports drink can, if you are an endurance athlete, mean better hydration[3] not only because you drink more fluid, but also because your body absorbs sports drinks slightly faster than water. And sports drinks replace carbohydrates that help fuel your muscles.

But you really only need the benefits of a sports drink during endurance exercise that lasts for more than 60 to 90 minutes. A recreational exerciser who works out for less than an hour has no need for sports drinks.

So remember: The taste of water may be simple, but its function isn't. Water is 100% natural, 100% pure, low sodium, calorie free, fat free, and cholesterol free—a practically perfect nutrient. Make sure you drink enough to stay hydrated for your health—and your athletic performance.

Remember: You, your physician, and your nutritionist need to work together to discuss nutrition concerns. The above information is not intended as a substitute for appropriate medical treatment.

References

1. Brouns F: Nutritional Needs of Athletes. England, Wiley & Sons, 1993, pp 50-69
2. Food and Nutrition Board, National Research Council. Recommended Dietary Allowances, ed 10. Washington DC, National Academy Press, 1989
3. Hubbard R, Szlyk R, Armstrong L: Influence of thirst and fluid palatability on fluid ingestion during exercise, in Gisolfi C, Lamb D (eds): Perspectives in Exercise Science and Sports Medicine, vol 3. Carmel, IN. Benchmark Press, 1990, pp 39-96 ∎

Questions

1. Compare how long a person can survive without food versus without water.

2. How does water help regulate body temperature?

3. Why is it dangerous to be dehydrated during exercise?

Answers begin on page 161.

47 *Iron is a trace mineral that the body needs for good health. It forms a critical part of the hemoglobin molecules in red blood cells. The reversible binding that takes place between hemoglobin and molecular oxygen occurs at the iron site. Public attention has often been focused on obtaining adequate amounts of this essential micronutrient in the diet or through mineral supplements. Now, however, recent evidence suggests that for middle-aged and older people, too much iron may be a more prevalent health concern than too little iron.*

Iron: Too Much of a Good Thing?

Once touted as a pep tonic, iron may actually increase the risk of deadly disease.

It wasn't all that long ago that radio and TV commercials hammered home the notion that older people could revitalize their "tired blood" by taking iron-rich *Geritol*, a remedy for what the ads implied was a virtual epidemic of iron-deficiency anemia. But anemia caused by consuming too little iron is far less common than those ads have led people to think. In fact, too much iron, not too little, may be a more prevalent health problem for middle-aged and older people—and a more ominous one as well. Recent evidence suggests that the extensive iron stores people accumulate as they age may increase the risk of cancer and coronary heart disease.

A Perilous Stockpile
Iron is an essential component of the red blood cells that haul oxygen through the blood; lack of iron causes anemia, or deficiency of red blood cells. But the vast majority of men and postmenopausal women consume more iron than they need.

Iron stores climb sharply during a man's early twenties, and continue to grow throughout life. By age 50, the average American man has accumulated enough iron, mainly in the spleen, bone marrow, and liver, to live for nearly four years without getting any additional iron from his diet. In women who haven't reached menopause, the monthly loss of iron-rich blood keeps iron stores from building up. But after menopause, they start catching up with men by building iron reserves even faster than men do.

All that stockpiled metal can cause trouble. As iron stores rise, so does the amount of free iron, metal that is not bound and inactivated by other molecules. That unbound iron speeds the creation of free radicals, highly unstable molecules that can contribute both to cancer and coronary disease.

Iron and Cancer
Studies in mice have shown that the growth rate of cancers depends directly on the amount of iron in the diet. Mounting evidence has linked excess iron with cancer in humans as well.

People who regularly donate blood, a practice that keeps their iron reserves down, have a decreased risk of cancer. Regular blood loss might also help explain the puzzling fact that premenopausal women have somewhat lower rates of most cancers than men of the same age.

One in every 250 Americans has hemochromatosis, a genetic disease that causes them to store huge amounts of iron. Researchers have long known that hemochromatosis boosts the risk of liver cancer roughly 200 times. In February, a Finnish study of more than 40,000 people showed that those who had much more iron than average were also three times more likely to develop colon or rectal cancer.

Two other studies published that same month have linked cancer to even moderate elevations in iron. Researchers sponsored by the U.S. and Japanese governments measured iron stores in more than 8500 American adults and then followed them for some 15 years. The 17 percent of men and 9 percent of women whose iron levels were clearly above average, but well below the levels found in people with hemochromatosis, had more than a 20 percent increased risk for all cancers combined.

In the second study, Illinois researchers evaluated some 360 people, including 145 who had precancerous colon polyps. After excluding those who might have

"Iron: Too Much of a Good Thing?" Copyright 1994 by Consumers Union of U.S., Inc., Yonkers, NY 10703-1057. Reprinted by permission from *Consumer Reports on Health*, July 1994.

had hemochromatosis, the researchers found that the risk of polyps increased steadily as iron stores rose; that risk was up to four times greater among people in the top half for iron stores compared with those in the bottom quarter.

Iron and Coronary Disease
Free radicals created by contact with iron may harm the heart in two ways. First, laboratory and animal research suggests that low-density lipoprotein (LDL) cholesterol, the "bad" kind, actually clogs arteries only after it has been oxidized by free radicals. In addition, the heart muscle may spill its stored iron if a blood-vessel spasm or clot interrupts its blood supply; when circulation resumes, that iron creates a horde of free radicals that may damage the muscle.

Two major clinical studies support those findings. Finnish researchers measured blood levels of ferritin, the protein that holds stored iron, in some 2000 men and then followed them for three years. The men who had ferritin levels in the high-normal range were more than twice as likely to have a heart attack as the men who had lower levels. Those who had high levels of both ferritin and LDL showed a striking five-fold increase in coronary risk, presumably because the iron helped oxidize the LDL.

In March, another large, carefully designed prospective study bolstered and refined the link between iron and coronary disease. Using a sophisticated questionnaire, Harvard researchers estimated the amount of iron consumed in the previous year by some 45,000 men. The researchers found no connection between overall iron intake and coronary risk. But not all iron is the same. When the body has enough iron, it sharply reduces its absorption of non-heme iron, the kind that comes from plants. But it keeps absorbing heme iron, the kind in meat, taking in up to 10 times more heme than non-heme iron.

Sure enough, the researchers found that intake of heme iron was directly related to coronary risk. Of course, heavy meat eaters consume large amounts not only of heme iron but also of saturated fat which raises LDL levels and might account for the increased risk. But the effects of heme iron persisted even after the researchers adjusted for saturated-fat consumption. The importance of heme iron may explain why the Finnish study did link overall iron intake with coronary disease: The Finns eat much more meat than Americans do—so most of the iron Finns consume is, in fact, heme iron.

(Several other studies have found no connection between iron levels and coronary risk. But none of those studies was designed nearly as rigorously as the Finnish and Harvard studies.)

What to Do
The notion that excess iron can jeopardize health needs to be evaluated in further studies. But the available evidence does warrant taking certain sensible precautions. First, avoid iron supplements unless you're pregnant or your doctor has determined that you have iron deficiency anemia. Second, ask your doctor about your red-blood-cell count a basic part of any routine blood analysis, next time you have a physical exam. (Consumers Union's medical consultants also recommend two related blood assays to screen for hemochromatosis: the serum-iron and transferrin-saturation tests. The tests, which total about $60, generally need to be done once every 5 years.) Unless your red-blood-cell count indicates low iron levels, you can then take the following steps to help stabilize or even reduce your iron levels:

Eat less meat, particularly red meat. Nutrition researchers currently recommend eating no more than two small, three-ounce servings of meat per day, mainly to limit fat consumption. Adhering to those guidelines or even cutting back further should help keep iron stores down.

Exercise regularly. Endurance athletes such as marathon runners have relatively low iron levels, apparently because the sustained movement causes minor gastrointestinal bleeding and destruction of red blood cells; that can even lead to "sports anemia." Moderate exercise poses no such danger in healthy people, since it causes much less bleeding and red-cell destruction. But working out regularly could actually cause nearly as much iron loss over the course of a month as menstruation does.

Consider giving blood. The ancient practice of frequent blood letting is now the established treatment for hemochromatosis. Less frequent blood donations may help lower iron levels in other people. Giving a pint of blood, for example, gets rid of more iron than a woman loses through menstruation over the course of an entire year. At the very least, giving blood occasionally won't do a healthy donor any harm, and it will do other people good. ■

Questions

1. Why can high stores of iron be harmful?

2. In what situations is taking iron supplements recommended?

3. What tips are given for stabilizing or reducing iron levels?

Answers begin on page 161.

48 *For years, science researchers have recognized the need to develop artificial blood to bolster the supplies in blood banks. Blood supplies, which play a critical role in saving lives, have a shelf life of only 28–42 days. With the AIDS epidemic, there is also the concern of viral contamination of supplies. The key step in making an artificial blood is finding a substitute for hemoglobin, the oxygen-carrying protein of blood. Polyheme, developed by Northfield Labs, may prove ideal.*

Artificial Blood May Be a Heartbeat Away

Ron Stodghill II

Northfield Labs is close to perfecting recycled hemoglobin.

Richard DeWoskin's quest began in 1970, as critically wounded American soldiers were bleeding on the battlefields of Vietnam. A young biology student at the time, DeWoskin was recruited by prominent Navy surgeon Dr. Gerald S. Moss to help develop artificial blood for the U.S. military.

It was the start of a rollercoaster ride from discovery and breakthrough to the brink of despair. Now the company they started, Northfield Laboratories Inc., may finally be nearing its goal: to create an inexpensive, abundant, and safe replacement for hemoglobin—the oxygen-carrying component in red blood cells. "Our work is finally paying off," says DeWoskin, now chairman and CEO. "We're not that far away."

A half-dozen other biotech and medical equipment companies—including giants Baxter International, Upjohn, and Eli Lilly—have targeted similar types of products. But recently, the Evanston (Ill.) startup has pulled ahead in all-important "Phase II" trials to establish efficacy and safety. If it can pass a final cluster of tests, Northfield's product, called PolyHeme, could be on the shelves in 1997.

Hypoglobin

A dose of skepticism couldn't hurt, however. Over the past decade, more hype than hemoglobin has issued from science labs in American, Europe, and Japan—and Northfield has been no exception. The fact is, human blood is fiendishly difficult to duplicate. And hemoglobin is the biggest challenge of all. This ironbearing protein, which makes blood red, binds oxygen and ferries it to tissues throughout the body. Massive blood loss quickly leads to oxygen deprivation, followed by shock, and failure of vital organs.

Physicians can bolster a patient's blood levels temporarily with salt or protein solutions that keep the heart pumping. But these can't deliver critical oxygen. Transfusing donated blood solves the problem but has clear disadvantages. It must be matched to the patient's blood type, which wastes precious time. It is prone to viral contamination, as the AIDS epidemic has demonstrated. And it has a short shelf life—just 28-42 days.

Consequently, blood is almost always in short supply. More than 4 million Americans receive upwards of 12 million units of it each year. But recently, supplies have dwindled because the AIDS menace, illogically, has scared away many potential donors. So-called blood substitutes, on the other hand, won't carry diseases or need to be cross-matched. And in theory, they can be stored safely for up to one year. As a result, the market for substitutes could run as high as $2.5 billion a year, according to analysts at Robertson Stephens & Co. in San Francisco.

Against Type

The stumbling block for scientists has been retrieving and isolating the 280 million hemoglobin molecules contained in each red blood cell so that they can be transfused no matter what a person's blood type is. Removed from a protective cell casing, however, the molecules tend to leak through the blood vessel wall into surrounding tissue. Unless they are genetically or chemically altered, the molecules can be toxic to nearly every organ in the body.

To get around that, companies have adopted several exotic approaches. Upjohn, in a partnership with Boston-based Biopure, is testing hemoglobin from cow's blood that is rejiggered to avoid attack by the body's immune system. Somatogen Inc. in Boulder, Colo., working with giant Eli Lilly & Co., is producing hemoglobin from genetically altered bacteria. Others hope to bypass nature's own mechanism entirely. San Diego-based Alliance, backed by Johnson & Johnson, is developing a substance based on nontoxic, oxygen-carrying chemicals called perfluorocarbons. Such substances have produced side effects, but Alliance has reformulated them to smooth the body's acceptance.

Northfield's approach is less technologically ambitious and may, therefore, be easier to perfect. Scientists begin by removing red blood cells from donated blood that has sat on a shelf longer than the time permitted by the Food & Drug Administration. Since expired blood has no other medical uses, Northfield can obtain it cheaply. Next, they extract hemoglobin molecules and link them together in long chains, called polymers. Their larger size prevents them from leaking from the blood vessels. The resulting solution is purified and bottled, giving it a long shelf-life.

This approach isn't entirely new. Northfield used similar techniques in the mid-1980s to produce its first experimental substitutes. Unfortunately, these caused breathing difficulties and muscular pains in some tests patients and had to be withdrawn from FDA trials in 1990.

Northfield says the problem was caused by naturally occurring contaminants the company had overlooked. With stepped-up purification, the symptoms disappeared. In March, Northfield received FDA approval to double its dosage of PolyHeme in ongoing human safety trials of 300 grams or six units—about three-fifths of the total blood in the human body—an unprecedented high level in the field of blood substitutes and about what might be required in a severe accident. Northfield's closet competitors, Baxter and Somatogen, have had difficulty moving beyond 30 grams in similar phase II trials. "Northfield has jumped way ahead," says Kevin C. Tang, a biotechnology analyst with Alex. Brown & Sons Inc. Investors are also pleased. The stock is trading at 13, double the price it listed at a year ago.

The doesn't mean the battle is over. "There are still a lot of fundamental questions that need to be answered," says Dr. Joseph C. Fratantoni, the FDA's chief medical reviewer for blood substitutes. Because of earlier safety problems—Northfield's and others'—the FDA remains wary of blood substitutes in general, and has yet to lay out specific endpoints for Phase III trials that establish efficacy and dosage. Says Fratantoni: "We've already got a product that's satisfactory in both safety and efficacy"—donated blood.

Trial Trials
Northfield's biggest problem is time. Even analysts who admire its prototype admit that Northfield may not have deep enough pockets for a prolonged Phase III trial. Unlike its larger and more diversified competitors, the company has no other products on the market or even in the pipeline. DeWoskin maintains that the need for a cash infusion is not immediate. The company has $12.5 million in the bank—sufficient to cover the next 20 months of expenses. "We're not strapped from an operating standpoint," he insists.

Expanded clinical trials and increased manufacturing will cost Northfield $25-$30 million, but DeWoskin and President Steven A. Gould believe they can sign up a partner—preferably a large pharmaceutical company that will trade cash and production capacity for PolyHeme marketing or distribution rights.

For DeWoskin, hashing out such deals is a minor detail compared with the enormity of proving the science—and helping patients. Since the 17th century, he notes, researchers seeking blood substitutes have tested everything from milk and vegetable oil to beer. If he can keep Northfield on track, they may be breaking out the champagne. ■

Questions

1. What are the disadvantages of transfusing donated blood?

2. Why can't hemoglobin be transfused as a blood substitute without its protective cell casing?

3. Describe Northfield's process for creating their blood substitute.

Answers begin on page 161.

Answers

1. **Metrics: Mismeasuring Consumer Demand**
   1. The International Bureau of Weights and Measures was established in 1875 by the Treaty of the Meter to provide metric standards of measurement for worldwide use.
   2. President Bush issued Executive Order 12770, which directed the Commerce Department to coordinate all federal agencies in converting to the metric system.
   3. Conversion to the metric system would enhance international trade and would allow standardized and simpler product packaging.

2. **Going Metric: American Foods and Drugs Measure Up**
   1. Only two: Liberia and Myanmar.
   2. Displaying both English and metric units on packaging will give consumers who are unfamiliar with the metric system a chance to learn about it.
   3. The metric system is easy to learn because everything is divisible or multipliable by factors of 10.

3. *Serendipity*
   1. Selenium in glass imparts a deep ruby coloration.
   2. Selenium improves the impact strength and machinability of steel.
   3. Selenium supplements help animals gain weight and prevent white muscle disease.

4. **Calcium: Here's How to Bone Up on This Essential Mineral**
   1. They are the same.
   2. The amount of calcium absorbed decreases as you age, especially after about age 65.
   3. Estrogen levels decrease after menopause, and estrogen helps to prevent bone loss.

5. **Gold: Noblest of the Nobles**
   1. Gold is found naturally in highly pure form and resists corrosion forever in ordinary conditions.
   2. Ductility, corrosion resistance, thermal and electrical conductivity.
   3. Computers, telephone jacks.

6. **Heavy-Ion Research Institute Explores Limits of Periodic Table**
   1. By the fusion of the nuclei of two lighter elements.
   2. Three.
   3. Inoperable cancers.

7. **The New Miracle Drug May Be—Smog?**
   1. Nitric oxide relaxes blood vessels and eases blood flow, keeping blood pressure in check.
   2. Nitric oxide boosts the body's ability to fight infections from microorganisms.
   3. It blocks the synthesis of NO by inhibiting enzymes.

8. **The Big Squeeze in the Lab: How Extreme Pressure Is Creating Exotic Materials**
   1. Diamond.
   2. It could serve as a dense form of stored chemical energy.
   3. Oxygen forms crystals that are no longer colorless.

9. **Chemical Techniques Help Conserve Artifacts Raised from *Titanic* Wreck**
   1. Exposure of artifacts to air accelerates corrosion, and leather items harden and crack when dried.
   2. Electrolysis loosens surface corrosion and removes from within the object chloride ions that might cause future corrosion.
   3. Glass.

10. Carbon Dioxide: Global Problem and Global Resource
1. The carbon dioxide buildup is suspected to be due to human-made emissions, particularly the burning of fossil fuels.
2. An advantage of the CO_2 process is that it avoids forming the toxic by-products associated with TiO_2 synthesis.
3. Carbon dioxide is abundant, it is cheap, and recycling CO_2 may save energy.

11. Physicists Create New State of Matter
1. To 180 billionths of a degree above absolute zero.
2. Lasers.
3. Albert Einstein, who based his theory on the work of Satyendra Nath Bose, an Indian physicist.

12. Tapping the Market for Bottled Baby Water
1. Infants get all the water they need from breast milk or formula.
2. Hyponatremia is a low concentration of sodium ions in the blood, which may be caused by consuming too much water.
3. Baby water is just water, while electrolyte solutions contain additional additives such as sodium and potassium ions and sugar to replenish the body's supplies following diarrhea.

13. Gas Hydrates Eyed as Future Energy Source
1. A molecule of methane encased in a spherical cluster of water molecules.
2. The water molecules are held together by hydrogen bonds.
3. Methane hydrate is a solid.

14. Dances with Molecules: Controlling Chemical Reactions with Laser Light
1. Using laser light to break specific chemical bonds would give control over the reaction and would enable a researcher to increase or decrease certain products.
2. Laser light can change the direction in which an electrical current flows in a semiconductor.
3. Lasers can be used to enhance drug synthesis, yielding a desired compound without unwanted side products.

15. Complexities of Ozone Loss Continue to Challenge Scientists
1. Balloons are launched at different times with instruments that probe the same air mass.
2. Chlorofluorocarbons and halons.
3. Polar stratospheric clouds (PSCs) provide surfaces for reactions that convert chlorine and bromine into active forms that can catalyze ozone destruction.

16. Drug Giants Ready to Enter Stomach-Medicine Battle
1. Antacids neutralize acid that is already present in the stomach.
2. Antacids start neutralizing acid the instant they enter the stomach. H2 blockers must be absorbed into the bloodstream before they will exert action.
3. Antacids last about two hours, while H2 blockers are effective for six to twelve hours.

17. Having the Last Gas
1. Landfill gas is 35–40% CO_2 and 60–65% methane.
2. Because volatile fatty acids are the principal products.
3. Methane can be explosive.

18. Making Molecular Filters More Reactive
1. Inorganic materials are inexpensive.
2. The size of the pores might be controlled.
3. The bonds that hold organic zeolites together are fragile.

19. The Chlorine Controversy
1. Chlorine is extremely reactive, and the resulting compounds are extremely stable.
2. Compounds that mimic estrogen have been linked with an increased risk of breast cancer.
3. Ozone dissipates more rapidly than chlorine, requiring more treatments. Ozone produces its own harmful products (carcinogens).

20. Wine and Heart Disease
1. High dairy fat consumption is known to increase coronary heart disease (CHD) mortality, but residents of some French cities had high fat consumption with low CHD mortality rates. Wine consumption explained the paradox.
2. Alcohol consumption decreased total cholesterol and increased the beneficial HDL level.
3. Antioxidants might block the oxidation of LDL, which is a key step in the development of arterial plaque.

21. Amoco, Haldor Topsoe Develop Dimethyl Ether as Alternative Diesel Fuel
1. Diesel fuel produces nitrogen oxides and soot.
2. DME has lower soot and nitrogen oxide emissions than diesel fuel, with comparable thermal efficiency.

3. Nitrogen oxide emissions contribute to acid rain, to the formation of ground-level ozone, and to the depletion of stratospheric ozone.

22. Clearing the Air
1. Mobile homes contain lots of new pressed-wood materials in a compact space.
2. Coat the wood products with a waterproof finish, such as polyurethane.
3. Air out the vehicle if it is new or has been in storage.

23. Pain, Pain Go Away
1. Coated aspirin may take up to twice as long to give relief compared to regular aspirin.
2. Acetaminophen is gentler on the stomach and decreases fever without the risk of Reye syndrome.
3. The OTC versions contain a lower dosage strength than the prescription versions.

24. Drugstore Deceptions: Separating Hype from Hope
1. Emotional stress has no effect on the body's need for vitamins.
2. Too much iron intake can mask serious underlying conditions and may increase the risk of cancer and heart disease.
3. Abnormal heart rhythms and increased blood pressure are associated with PPA.

25. Caffeine: Grounds for Concern?
1. It will double their chance of miscarriage.
2. The response to caffeine varies considerably from person to person.
3. Polyphenols.

26. Green Chemistry at Work
1. Vanillin (used in ice cream), hydroquinone (used in photography), nylon 66 (a synthetic fiber), and L-DOPA (a drug for treating Parkinson's disease).
2. It is a carcinogen.
3. The key is the use of biocatalysts (enzymes) in the various processes.

27. What's Wrong with Sugar?
1. The study concluded that sugar didn't cause hyperactivity and even moderately relaxed some of the children.
2. An increased sugar level causes more insulin secretion by the pancreas, which acts to draw sugar out of the bloodstream.
3. Aspartame is unstable at high temperatures.

28. Fiber Bounces Back
1. It inactivates certain digestive acids made from cholesterol in the liver, and in response the liver draws cholesterol out of the blood in order to make more acids.
2. Soluble fiber helps decrease blood glucose by slowing absorption of sugar from the small intestine.
3. Fiber speeds potentially cancer-causing wastes through the colon and also increases stool size, thus decreasing the concentration of those wastes.

29. New Oils for Old
1. Plants.
2. Oleochemical-derived lubricants might be the best choice when the lubricant is lost to the environment (i.e., chain-saw lubricant) because these lubricants are biodegradable.
3. Petroleum-based inks contain carcinogens and are difficult to recycle because harmful solvents are needed to remove them from paper. Oleochemical-based inks are nontoxic and can be removed enzymatically from paper.

30. Estrogen: Friend or Foe?
1. To prevent pregnancy, to relieve symptoms of menopause, to prevent osteoporosis, or to treat some forms of infertility and menstrual disorders.
2. DES was used to prevent miscarriage. Children of DES mothers have a higher incidence of reproductive tract abnormalities and may be at increased risk of developing some types of cancer.
3. None. They are not subject to premarket clearance and do not have to provide the FDA with proof of safety.

31. Protein Devices May Increase Computer Speed and Memory
1. Bacteriorhodopsin.
2. One thousand times more information.
3. The cleansing and etching steps in the production of semiconductors involve a major amount of pollution, whereas isolating bacteriorhodopsin is environmentally safe.

32. Designer Proteins: Building Machines of Life from Scratch
1. Protein shape in water is driven by the efforts of the protein's water-repelling side chains to escape.
2. Specially placed disulfide bonds hold it together.
3. The hypothesis is that the protein's amino acid sequence must contain hydrophilic and hydrophobic sections in very specific locations in order to fold properly. The details of which amino acid goes where and the order are less important.

33. Novo Nordisk's Mean Green Machine
1. Tide contains Carezyme, a proprietary enzyme that decreases fuzz buildup on fabrics.
2. Hexane is explosive and dangerous to breathe. Novo's enzyme decreases risks to workers and to the environment.
3. Novo transferred the gene for producing lipolase enzyme to another microbe that was able to produce more of the enzyme.

34. Better by Design: Biocatalysts for the Future
1. There is a high cost in producing enzymes pure enough to obtain appropriate specificity, efficiency, and safety for food processing.
2. Liposomes help stabilize the enzymes, give uniform enzyme distribution, and speed up ripening of cheese.
3. Because of the inability to expose the cellulose in the wood to attack by the cellulases.

35. Life in Boiling Water
1. They are able to live in extreme conditions (high temperature, high salt, low pH).
2. Hyperthermophiles reduce sulfur to sulfide during growth, creating highly corrosive and toxic mixtures as well as an unpleasant working environment.
3. With the thermostable DNA polymerase, there is no need to add more enzyme after each cycle. It also produces higher fidelity copies of DNA.

36. The Gene Kings
1. The cDNA is more sturdy than the fragile mRNA.
2. Merck hoped to undermine those people who are filing patents in an attempt to corner the gene market.
3. Between 85 and 90%.

37. Environmental Risks in Agricultural Biotechnology
1. Some people believe that biotechnologists should not be "playing God."
2. Bromoxynil increases crop yields, but it also kills mammals.
3. Bt endotoxins give plants a tolerance toward pests, but the endotoxins might also kill soil fauna.

38. Transgenic Livestock in Agriculture and Medicine
1. Combinations of regulatory elements and structural genes.
2. Most of the offspring do not incorporate the new genes into their system.
3. Contamination with human retroviruses, such as HIV, is avoided.

39. Cancer-fighting Foods: Green Revolution
1. Functional foods are those with ingredients thought to prevent disease.
2. Phytochemicals help plants resist oxidation, viral attack, harsh weather, and the insults of handling.
3. Isoflavons are found in soy products. They act as antioxidants, carcinogen blockers, and tumor suppressors.

40. B Makes the Grade
1. Folic acid prevents birth defects, including spina bifida and anencephaly.
2. High homocysteine levels increase the risk of heart attack. B vitamins decrease homocysteine levels.
3. The body needs them to manufacture neurotransmitters, chemicals that facilitate nerve signals in the brain.

41. Hypoglycemia: Fact and Fiction about Blood Sugar
1. Symptoms include headache, fatigue, tremulousness, sweating, palpitations, and inability to concentrate several hours after a high-carbohydrate meal.
2. Patients eat foods that normally cause symptoms and then visit the doctor's office to have blood drawn and tested.
3. Liver disorders, insulin-producing tumors, adrenal gland failure, insulin and oral diabetes drugs, and stomach surgery can disrupt the glucose-insulin balance.

42. Marketers Milk Misconceptions on Lactose Intolerance
1. Lactose maldigestion is not necessarily accompanied by unpleasant GI side effects.
2. Irritable bowel disease, parasitic infection, and stomach flu.
3. The higher the rise in blood-sugar level, the more lactose that has been digested and therefore the less likely the person is lactose intolerant.

43. Nutrition: Pasta Is Not Poison
1. Only if a person eats too much of them.
2. People are obese when they weigh at least 20% more than they should.
3. In insulin resistance, cells won't permit glucose entry unless there's an abnormally large quantity of insulin on hand to open the door. Some people are genetically predisposed to insulin resistance; others develop the condi-

tion as a consequence of being obese and sedentary.

44. The New Skinny on Fat
1. By diminishing appetite and speeding up metabolism.
2. A strain of obese mice was discovered in the 1950s. Researchers believed the mice were missing a particular gene, the ob gene, that lets the brain know how much fat is in storage. In December 1994, researchers identified the gene and its hormone.
3. After animal testing to make sure it's safe.

45. Cutting Cholesterol: More Vital Than Ever
1. In men, decreased cholesterol levels were linked to a decreased overall death rate. In women, decreased cholesterol levels were linked to a decreased risk of a fatal heart attack.
2. Plaques became more smooth and stable, thus decreasing the risk of plaques breaking open and clots forming, leading to a heart attack.
3. Lose weight to increase HDL and decrease LDL; minimize consumption of saturated (animal) fats; exercise; and stop smoking.

46. Water: The Ultimate Nutrient
1. A person can survive weeks without food, but only a few days without water.
2. Water absorbs heat from the muscles and transports it to the skin's surface.
3. Dehydration reduces blood volume, which slows down the transport of chemicals to and from the muscles.

47. Iron: Too Much of a Good Thing?
1. As iron levels rise, unbound iron increases, leading to the formation of free radicals which can contribute to coronary diseases and cancer.
2. Iron supplements are fine for pregnant women or people with iron-deficiency anemia.
3. Eat less meat, particularly red meat; exercise regularly; consider giving blood.

48. Artificial Blood May Be a Heartbeat Away
1. Blood types must match; donated blood is prone to viral contamination; donated blood has a short shelf life.
2. Hemoglobin would leak through blood vessel walls into surrounding tissues.
3. They remove red blood cells from expired donated blood and extract the hemoglobin. The hemoglobin molecules are then linked together to form a polymer.